T0295153

Repair of Advanced Composites for Aerospace Applications

Repair of Advanced Composites for Aerospace Applications

Edited by
Mohamed Thariq Hameed Sultan
Murugan Rajesh
Kandasamy Jayakrishna

CRC Press
Taylor & Francis Group
Boca Raton London New York

CRC Press is an imprint of the
Taylor & Francis Group, an **Informa** business

First edition published 2022
by CRC Press
6000 Broken Sound Parkway NW, Suite 300, Boca Raton, FL 33487-2742

and by CRC Press
2 Park Square, Milton Park, Abingdon, Oxon, OX14 4RN

Library of Congress Cataloging-in-Publication Data
Names: Thariq, Mohamed, editor. I Rajesh, M. (Mechanical engineer), editor. I Jayakrishna, K., 1984- editor.
Title: Repair of advanced composites for aerospace applications /
edited by Mohamed Thariq Hameed Sultan, M. Rajesh, K. Jayakrishna.
Description: First edition. I Boca Raton : CRC Press, 2022. I Includes bibliographical
references and index.
Identifiers: LCCN 2021049768 (print) I LCCN 2021049769 (ebook) I ISBN 9781032061634 (hardback) I
ISBN 9781032061641 (paperback) I ISBN 9781003200994 (ebook)
Subjects: LCSH: Polymeric composites. I Airplanes–Materials. I Airplanes–Maintenance and repair. I
Space vehicles–Materials. I Space vehicles–Maintenance and repair.
Classification: LCC TA418.9.C6 R455 2022 (print) I LCC TA418.9.C6 (ebook) I
DDC 620.1/92–dc23/eng/20211118
LC record available at https://lccn.loc.gov/2021049768
LC ebook record available at https://lccn.loc.gov/2021049769

ISBN: 978-1-032-06163-4 (hbk)
ISBN: 978-1-032-06164-1 (pbk)
ISBN: 978-1-003-20099-4 (ebk)

DOI: 10.1201/9781003200994

Typeset in Times
by SPi Technologies India Pvt Ltd (Straive)

Contents

Preface

Lightweight polymer composites are highly preferred for weight-sensitive applications, especially in aerospace, to drastically bring down the structure's final weight and fuel consumption. However, lightweight composites in the aerospace industry experience unexpected impacts due to service load, bird strikes during flight operations and dropping of hand tools during maintenance work. The composite can be damaged due to matrix cracking, fibre fracture, fibre pull-out and delamination, thus reducing the strength and stiffness properties of the composite. Hence it is essential to enhance resistance against damage in the lightweight composites through effective reinforcement and the manufacturing process. In this book, an insight into the repair of polymer composites for aerospace applications is discussed, with a a focus on factors affecting the performance of the lightweight composites, fibre failure modes, fibre failure and its propagation. The book also focuses on the characterization of polymer composites and the effect of reinforcement through simulation of the composites' damage mechanisms concerning specific environments and applications. This book covers versatile topics such as Natural Fibre Reinforced Polymer (NFRP) composites, types of bonding, manufacturing techniques of composite joint fabrication and adhesive strength, effect of temperature on deflection behaviour and repair techniques of cracking on failure of composite types. It also covers the importance of the vibration behaviour of the composite and explains in detail how the composites' fundamental natural frequency and corresponding damping factor are affected by the service load.

The editors would like to thank all authors for sharing their valuable findings, ideas and knowledge on the repair of composites. We are grateful to all reviewers for their valuable comments in improving this book. We also thank the publishing team for their continuous support and cooperation in bringing this book.

Editors:
Mohamed Thariq Hameed Sultan
Murugan Rajesh
Kandasamy Jayakrishna

Editors' Biographies

Prof. Ir. Ts. Dr. Mohamed Thariq Hameed Sultan completed his PhD in 2011 from the Department of Mechanical Engineering, University of Sheffield, United Kingdom in the field of mechanical engineering. He specializes in the field of hybrid composites, advance materials, structural health monitoring and impact studies. He is also a professional engineer (PEng) registered under the Board of Engineers Malaysia (BEM) and a charted engineer (CEng) registered with the Institution of Mechanical Engineers (IMechE) United Kingdom. Recently, he was awarded Professional Technologist (PTech) from Malaysian Board of Technologists (MBOT). He has published more than 220 journal articles and 15 books internationally.

Prof. Murugan Rajesh completed his PhD in 2017 from the Department of Mechanical Engineering, National Institute of Technology Karnataka, Surathkal, Mangalore, India in the field of composite material. Currently he is working in the School of Mechanical Engineering at the Vellore Institute of Technology University, India. He specializes in the field of hybrid composites, advance materials, composite joint and its failure analysis. He has published more than 15 journal articles in leading SCI Indexed journals, 15 book chapters and one book internationally.

Prof. Kandasamy Jayakrishna is an associate professor in the School of Mechanical Engineering at the Vellore Institute of Technology University, India. Dr. Jayakrishna's research is focused on bio-composites and management of manufacturing systems and supply chains to enhance efficiency, productivity and sustainability performance. More recent research is in the area of developing tools and techniques to enable value creation through sustainable manufacturing, including methods to facilitate more sustainable product design for closed-loop material flow in industrial symbiotic setup, and developing sustainable products using hybrid bio composites. He has published 52 articles in leading SCI/ SCOPUS Indexed journals, 23 book chapters, 90 refereed conference proceedings, and 1 book in the CRC/Springer Series. His initiatives to improve teaching effectiveness have been recognized through national awards. He has also been awarded Institution of Engineers (India) – Young Engineer Award in 2019 and Distinguished Researcher Award in the field of sustainable systems engineering in 2019 by the International Institute of Organized Research.

Contributors

Mohamad Ridzuan Ahmad
Industrial Technology Division
Malaysian Nuclear Agency
Bangi, Malaysia

Siti Madiha Muhammad Amir
Department of Aerospace Engineering,
  Faculty of Engineering
Universiti Putra Malaysia
Serdang, Malaysia
  and
Industrial Technology Division
Malaysian Nuclear Agency
Bangi, Malaysia

Ahmad Hamdan Ariffin
Faculty of Mechanical and
  Manufacturing Engineering
Universiti Tun Hussein Onn Malaysia
  (UTHM)
Batu Pahat, Malaysia

Ananda Babu A
School of Mechanical Engineering
Sharda University
Greater Noida, India

Saurabh Bait
Mechanical Engineering Department
Marathwada Mitra Mandal's Institute
  of Technology
Pune, India

Amol Bhanage
Mechanical Engineering Department
Marathwada Mitra Mandal's Institute
  of Technology
Pune, India

Thulasidhas Dhilipkumar
School of Mechanical Engineering,
Vellore Institute of Technology
Vellore, India

Muthukumaran Gunasegeran
School of Mechanical Engineering,
Vellore Institute of Technology
Vellore, India

Syed Noor Hussain bin Mohamed
  Hasanuthin
Faculty of Mechanical and
  Manufacturing Engineering
Universiti Tun Hussein Onn Malaysia
  (UTHM)
Batu Pahat, Malaysia

Tayfun Inik
Department of Mechanical Engineering
Zonguldak Bulent Ecevit University
Zonguldak, Turkey

A R Irfan
Fakulti Teknologi Kejuruteraan
  Mekanikal
Universiti Malaysia Perlis (UniMAP)
Pauh Putra Arau, Malaysia

Muhamad Noor Izwan Ishak
Industrial Technology Division
Malaysian Nuclear Agency
Bangi, Malaysia

Mohit Kumar Jain
School of Mechanical Engineering,
Vellore Institute of Technology
Vellore, India

Mohammad Jawaid
Laboratory of Biocomposite
  Technology
Institute of Tropical Forestry and Forest
  Products (INTROP), Universiti Putra
  Malaysia,
Serdang, Malaysia

Anita Jessie
School of Civil Engineering
Vellore Institute of Technology
Vellore, India

Tanay Kuclourya
School of Mechanical Engineering,
Vellore Institute of Technology
Vellore, India

Lakshmi Narayanan V
School of Mechanical Engineering
Vellore Institute of Technology,
Vellore, India

Mohamad Norani Mansor
Faculty of Mechanical and
  Manufacturing Engineering
Universiti Tun Hussein Onn Malaysia
  (UTHM)
Batu Pahat, Malaysia

Shukri Mohd
Industrial Technology Division
Malaysian Nuclear Agency
Bangi, Malaysia

Shubham Mudliar
School of Mechanical Engineering,
Vellore Institute of Technology
Vellore, India

Murugan Rajesh
School of Mechanical Engineering,
Vellore Institute of Technology
Vellore, India

Mohammad Sukri Mustapa
Faculty of Mechanical and
  Manufacturing Engineering
Universiti Tun Hussein Onn Malaysia
  (UTHM)
Batu Pahat, Malaysia

Natesh M
Department of Mechanical Engineering
V.S.B. Engineering College,
Karur, India

Syafiqah Nur Azrie Safri
Laboratory of Biocomposite
  Technology
Institute of Tropical Forestry and Forest
  Products (INTROP), Universiti Putra
  Malaysia,
Serdang, Malaysia

Ramesh Sakhare
Mechanical Engineering Department
Marathwada Mitra Mandal's Institute of
  Technology
Pune, India

Khairul Anuar Mohd Salleh
Industrial Technology Division
Malaysian Nuclear Agency
Bangi, Malaysia

Suhairy Sani
Industrial Technology Division
Malaysian Nuclear Agency
Bangi, Malaysia

Ain Umaira Md Shah
Department of Aerospace Engineering,
  Faculty of Engineering
Universiti Putra Malaysia
Serdang, Malaysia

Sourabh Kumar Soni
School of Mechanical Engineering
Vellore Institute of Technology,
Vellore, India

Soundhar A
Department of Mechanical Engineering
Sri Venkateswara College of
  Engineering (SVCE),
Chennai, India

Edwin Sudhagar P
School of Mechanical Engineering,
Vellore Institute of Technology
Vellore, India

Mohamed Thariq Hameed Sultan
Department of Aerospace Engineering,
    Faculty of Engineering
Universiti Putra Malaysia
Serdang, Malaysia

Laboratory of Biocomposite
    Technology
Institute of Tropical Forestry and Forest
    Products (INTROP), Universiti Putra
    Malaysia,

Serdang, Malaysia
Aerospace Malaysia Innovation Centre
    (944751-A)
Prime Minister's Department,
Cyberjaya, Malaysia

K. Tabreja
Department of Aerospace Engineering,
    Faculty of Engineering
Universiti Putra Malaysia,
Serdang, Malaysia

Narendiranath Babu Thamba
School of Mechanical Engineering,
Vellore Institute of Technology
Vellore, India

Benedict Thomas
School of Mechanical Engineering
Vellore Institute of Technology,
Vellore, India

Venkatesan S
School of Mechanical Engineering,
Vellore Institute of Technology
Vellore, India

Mehmet Yetmez
Department of Mechanical Engineering
Zonguldak Bulent Ecevit University
Zonguldak, Turkey

1 A Review on Polymer Nanocomposites Used in Aerospace

Benedict Thomas and Sourabh Kumar Soni
Vellore Institute of Technology, Vellore, India

CONTENTS

1.1 INTRODUCTION

Polymers are the most promising materials owing to their reproducibility and ease of processing. They are relatively inexpensive and widely employed in numerous engineering applications. In earlier literature, polymers are classified as (1) thermoplastics, (2) thermosets and (3) elastomers. Amalgamation of two or more materials such as polymer matrices and additives are combined to fabricate polymer composites (PMCs). There are some limitations in optimizing the characteristics of PMCs by incorporating traditional micron-scale fillers. Generally the conventional filler content in PMCs in the range of 10–70 wt% makes the resulting composite more expensive and dense. Unlike traditional PMCs comprising micro-scale fillers, the addition of nanoscale reinforcement materials (nanofillers) into a polymer creates a very short distance among the fillers; therefore, the characteristics of the resulting composites can be largely improved even with an enormously trivial nanofiller content. Owing to those improved characteristics, PMCs reinforced with nano-sized fillers are an open door as future materials in nanostructures, biomechanics, aviation and more.

During the last twenty years improvement in the functionalities of PMCs with nano-scaled fillers has come to be an attractive new subject in the field of nanocomposites, literature shows, and still progressive development is going on in this field to fill the lacuna (Wernik and Meguid 2010).

In earlier literature, PMCs reinforced with carbon nanotubes (CNTs) – also known as polymer nanocomposites (PNCs) – are more popular and widely studied among nanocomposite researchers. After Ajayan et al. (1994), Qian and their co-workers (Khalili and Haghbin 2013) were the initiators to examine the influence of the addition of CNTs in polymers by conducting experimental investigations. Numerous review articles have been published on the fabrication technique (Spitalsky et al. 2010; Jin and Park 2011), mechanical characterization (Spitalsky et al. 2010; Coleman et al. 2006) and thermal conductivities (H. Chen et al. 2016) of PNCs. Findings from published articles concur that the higher interfacial bond augments the mechanical characteristics because the CNTs take a major part of the load, which is externally applied in the PNCs. Moreover, CNTs are often modified (functionalised) to increase their dispersibility and facilitate their improved interactions with the polymers (Soni et al. 2020). Available literatures suggest that functionalization processes such as oxidation, amino-functionalization, polymer grafting and silane-functionalization are a promising approach that develops covalent interaction and integrates CNTs into the polymer matrix. Owing to that, resulting nanocomposites become more uniform and don't indicate any phase separation and agglomeration difficulty (Rahmat and Hubert 2011). In various research work physical and chemical functionalization of CNTs has been performed to increase interfacial adhesion; the same has been proven through an extensive experimental and analytical approach.

Some of the researchers observed enhancement in load transfer capability and mechanical properties, mainly due to chemical bonding among the treated CNTs and polymeric matrix (Liao and Li 2001). Pande et al. 2009 observed augmentation in the mechanical characteristics of MWCNT-reinforced PMMA polymer nanocomposites (PNCs). They also found that even trivial loading of functionalised CNTs in a PMMA matrix provide higher strength and modulus by virtue of better interaction among the reinforcement and matrix. Further, Ma et al. (2010) revealed that the homogenous exfoliation of nanotubes within the matrix depends considerably on numerous parameters such as CNT length, their entanglement phenomenon, V_{CNT}, viscosity of the matrix, attraction between inter-tube and dispersion methods. They also observed that the numerous CNT functionalization techniques and dispersion techniques were adopted by previous researchers to attain homogenous dispersion of nanotubes on the PMCs. After intensive review, Kim et al. (2012) found that uniform distribution in various solvents or polymer matrices is a major concern for the application of CNTs in various fieldws. Further, they presented various dispersion techniques and their effectiveness during the fabrication of nanocomposites. Following the morphological investigation of CNT/epoxy PNCs, Gojny et al. (2003) found that by using chemical functionalization techniques they effortlessly augment the dispersal of MWCNTs on the epoxy. Their findings showed enhanced interfacial interaction among the matrix and functionalised nanotubes, which results in improved mechanical characteristics of the resulting nanocomposite. Yu et al. (2006) prepared SWNT/epoxy PNC by utilising raw and purified SWNTs and conducted experimental studies to explore the influence of SWCNTs' purity on the thermal conductivities of SW-CNTRCs. They found that purified and functionalised SWCNTs–reinforced composites have five times better thermal conductivity as compared to unpurified reinforced composites. Also, as compared to end-walled and side-walled covalent chemically modified CNTs, Spitalsky et al. (2009) found stronger molecular interaction among the physically modified CNTs and polymer matrix. More detailed investigations revealed that polymerization is also an efficient approach; however, its efficiency is limited to specific polymers. In order to overcome the preceding problem and achieve the homogenous dispersion of CNTs with good alignment in numerous polymer matrices, alignment techniques such as mechanical stretching, infiltration winding, melt spinning, dielectrophoresis, application of electrical or magnetic fields, and ultrasonic-assisted solution evaporation have been also employed during the preparation of PNCs (Bhattacharya 2016).

1.2 RESEARCH WORKS ON THE EXPLORATION OF CNTRCs CHARACTERISTICS

In this section, an effort has been made to review published literature on the mechanical, thermal, thermo-mechanical and vibrational properties of CNT-reinforced PMCs. In addition, some literatures on the agglomeration effect of CNTs on composites reinforced with CNTs is also presented.

1.2.1 MECHANICAL CHARACTERISTICS OF CNTRCs

A substantial amount of experimental and numerical investigation have been carried out to analyse the mechanical characteristics of CNT-reinforced PMCs (CNTRCs). The main aim of this section is to summarise several destructive and non-destructive approaches for evaluation of mechanical properties and to present predictive strategies taken by numerous researchers to achieve the maximum advantage of CNTs reinforcement in numerous CNTs–based nanocomposites. In the preceding twenty years much intensive research on the fabrication and evaluation of the mechanical properties of numerous PNCs has been conducted and a large volume of the literatures suggests that the PNCs' mechanical performances are significantly reliant on the interfacial interaction among the CNTs and the polymers. Exploration of PS /MWCNTs nanocomposites were performed by Qian et al. (2000). Their observation revealed that upon loading of 1 wt% MWCNTs on PS /MWCNTs nanocomposites, improvement in break stresses (25%) and E (36%–42%) were achieved as compared to blank PS composites, primarily due to homogenous dispersion of undamaged NTs on PS matrices. They also performed microstructural investigation, which shows external loads are effectively transferred to NTs due to their lesser size and larger l/d ratios.

Further, Safadi et al. (2002) conducted fabrication and orientation processes of a PS/MWCNTs composite by using spin casting and film casting techniques to investigate the basic relations between the processing conditions and their mechanical and electrical properties. Uniform dispersion of MWCNTs in solutions was achieved by a simple sonication process, and the MWCNTs' orientation and dispersion within a PS matrix were inspected employing SEM. The researchers found higher orientation of MWCNTs near the fluid-solid interface and decreases to random orientation at the shear-free fluid-air interface. By using the Carreau equation, rheology of PS/MWNTs suspensions was modeled. Their results shows an increment in the tensile strength (TS) of the nanocomposite, which is loaded with up to 1 vol% MWCNTs. Numerous techniques were employed in literatures to fabricate CNT–reinforced polyvinyl alcohol (PVA) composites such as gel spinning, electro spinning, wet spinning (Jee et al. 2012) and melt spinning (Ferreira et al. 2017). Yang et al. (2015) prepared the PVA/ MWCNT composites fibre by using the melt-spinning process to investigate the mechanical response and microstructure characteristics as a function of the draw condition. They found higher elastic and mechanical properties as compared to undrawn composite fiber; XRD analysis also revealed that by increasing the draw ratio the degree of orientation and crystallinity increases. Polypropylene (PP) thermoplastics are also a widely employed polymer owing to its low density and ease of processability as well as a decent balance among cost and properties. Manchado et al. (2005) analysed the effect of numerous SWCNTs concentrations on physical and thermal characteristics of CNT–reinforced PMCs by using DMA and calorimetric analysis. In their fabrication procedure, a shear mixing approach was used to attain homogenous dispersion of SWCNTs in isotactic polypropylene (iPP). Their results show that modulus of elasticity (E) and TS enhances with the accumulation of NTs, with a maximum content of 0.75 wt%. Also, the accumulation of a trivial amount of SWNT (less than 1 wt%) increases the rate of polymer crystallization with no considerable variations in the crystalline structure.

Epoxy is costly compared to other polymers, but due to its superior characteristics it is widely used in a range of industries, including aerospace. Earlier literature shows that epoxy is the most-studied and most-used polymer in advanced PNCs because of its prodigious properties. The foremost reason behind its popularity is its low viscosity, excellent thermo-mechanical properties, higher strength, low shrink rates, low volatility during cure and low flow rates. While investigating properties of MWCNTs/ epoxy PNCs, Montazeri et al. (2010) found that acid-treated MWCNTs are uniformly dispersed as compared to untreated MWCNTs within the nanocomposites. They also perceived that the homogenous exfoliation of nanotubes (MWs) in the epoxy increases the uniformity of stress distribution and diminishes stress concentration. While investigating mechanical characteristics of PNCs, Guo et al. (2007) found augmentation in the TS and fracture strain of epoxy-based PNCs. Montazeri et al. (2011) also observed significant effect of nanotubes on the PMCs involving the viscoelastic and mechanical characteristics of the MWCNT/epoxy PNCs. Mansour et al. (2013) analyses the mechanical response of the MW-CNTRCs and found good agreement between the value of elastic modulus estimated by the nano-indentation technique (NT) and tensile testing. They found that the both hardness and modulus of MWCNTs/epoxy composites increase with higher MWCNTs concentrations.

1.2.2 Thermal Characteristics of CNTRCs

Presently, thermal characteristics are the most important parameter for upcoming technologies. The excellent thermal conductivity and thermal stability of CNTs motivated substantial research interest to use them as a filler material and enhance

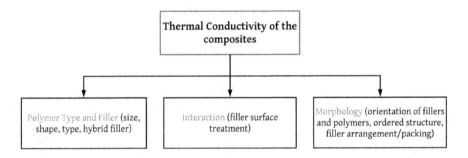

FIGURE 1.1 Factors that affects the thermal conductivities of composites. (Chen et al. 2016).

the thermal characteristics of polymers. Earlier literature shows that the thermal characteristics of PNCs depends on various aspects such as CNT content, uniform dispersion, aspect ratio and interfacial interactions with matrix, as shown in Figure 1.1.

Biercuk et al. (2002) suggested that the addition of trivial amounts of SWCNTs (1%) can enhance the thermal characteristics of industrial epoxy (up to 120%) and hence does not require any chemical functionalization process. Their initial findings show that CNTs are a tremendous filler material for producing nanocomposites with greater thermal conductance properties. A similar observation was realised by Velasco-Santos et al. (2003), who found that a trivial amount of CNTs in a methyl-ethyl methacrylate copolymer (MEMA) polymer matrix is efficient to enhance the storage modulus and thermal stability. Xiong et al. (2006) examined the microstructure of PU-MWCNT composite by using field-emission (FE)-SEM and TEM. Thermal and mechanical properties of the composite are estimated by DMTA, TGA and tensile testing. They observed excellent microstructure properties and increased T_g as well as thermal stability and increased value of TS by incorporating CNT into the matrix. Bin-Jie et al. (2013) investigated the thermal conductivity and stability of the fabricated PSA-MWCNT nanocomposite specimens by varying the CNTs content. They presented thermal response of PSA-MWCNT composites by using TGA; additionally they describe the composites' thermal conductive behaviour by using a heat conducting model. Concluding remarks presented by Wang et al. (2009) unearthed that the thermal conductivity of composites comprising short CNTs is greater than those having pristine CNTs. They fabricated PNCs in which mechanically shortened CNTs were employed as a reinforcement. Further, SEM and AFM characterizations suggested that dispersion of CNTs across the matrix is enhanced by NTs shortening. Kwon et al. (2013) investigated the thermal characteristics of PDMS-MWCNT PNCs as a function of V_{CNT} by employing a steady-state measurement (GHP) technique. They found notable augmentation in the thermal conductivity (increment of 390%) by the accumulation of trivial content (1.4 vol%) of MWCNTs in the composite.

1.2.3 RESEARCH WORKS ON EXPLORATION OF THERMO-MECHANICAL CHARACTERISTICS OF CNTRCs

Fidelus et al. (2005) explored the thermo-mechanical characteristics of a PNCs–comprising matrix as an epoxy and reinforcement as a randomly oriented SW and MWCNTs. Their thermo-mechanical properties were measured by using DMTA and an impact-testing technique. Their quantitative results show modest enhancements in tensile impact toughness and Young's modulus (E) of specific nanocomposites with the accumulation of trivial content of CNTs as compared to pure matrix material. Additionally, a somewhat higher elastic modulus of SWCNT–based nanocomposite was observed as compared to the value projected by the Krenchel model for short-fiber composites with random orientation. Montazeri et al. (2011) explored the tensile and thermo-mechanical characteristics of the epoxy-MWCNT nanocomposites by conducting DMTA and tensile tests investigations. The 0.1, 0.5 and 1.5 wt% of CNTs (MW) were assorted with epoxy at $80°$ C. Their presented results show that the reinforcement of 0.5 wt% MWCNTs was promising in improving the storage modulus value, whereas the specimen comprising 1.5 wt% MWCNTs showed the increment of 20.6% and 15.6% in the TS and modulus values. For the uniform dispersion of MWCNTs within a commercial epoxy resin, Martone et al. (2012) adopted solvent-aided dispersion and direct mixing dispersion techniques. Further to explore the effect of these two techniques on the microstructure of the resulting PNCs, morphological investigations were conducted. Based on the DMA and TMA results, they revealed that the probe sonication was promising for exfoliation and mixing.

Initially Gaidukovs et al. (2016) prepared graphite oxide (GO) and mixed it with CNTs by using ultrasonic treatment to fabricate polyurethane (PU) composites. Further thermo-mechanical characterization of the CNT/GO composites were conducted by employing TMA and DMA measurements, then compared with the unfilled PU. They observed enhancement in the storage moduli and loss moduli of the prepared PNCs as compared to the unfilled PU. Their concluding remarks suggest that the thermo-mechanical characteristics of the PU composites were considerably influenced by the accumulation of CNT/GO filler; increases in T_g were also observed due to strong polymer-filler interfacial interaction. Finally, their in-depth investigation shows that the most significant factor for the improved thermo-mechanical properties is temperature and the content of the filler. Sharma et al. (2015) then studied the thermo-mechanical characteristics of the MWCNT/polycarbonate (PC) composite employing MD simulation techniques. They found that the reinforcement of 2% CNTs (MW type) in the matrix is enough, promising to achieve PNCs with both static as well as dynamic characteristics. Their obtained results revealed that the sufficient load transfer capacity of nanotubes, greater surface area and interfacial interaction among the PC matrix and MWCNTs reinforcement are the responsible parameter for enhancement of the mechanical characteristics.

1.2.4 RESEARCH WORKS ON EXPLORATION OF VIBRATIONAL CHARACTERISTICS OF PMCs REINFORCED WITH CNTs

In this section, an attempt has been made to review the earlier literature on the vibrational response of the PMCs reinforced with CNTs. If any component or structure of mechanical systems is subjected to internal or external loads, then due to fluctuations on forces,energy vibration is induced, which is a major concern for the researchers. That vibration creates difficulties in operation, loss in energy, inaccuracies in performance, fatigue and even failure of the component or structure. Various intensive vibration analyses of the structure revealed that a large amount of vibration create a fluctuation in stiffness, mass and damping responses and changes the natural frequencies and mode shapes of the structures. Numerous in-depth investigations from renowned researchers also suggested that CNTs are great aspirants for the potential applications in improvement of energy absorption capability of composites.

Li et al. (2015) studied the vibration response of cement composite specimens with numerous CNT loadings by using the logarithmic decrement method and DMA. The influence of CNTs on the morphology was examined by employing MIP and SEM, which revealed that by incorporation of CNT in the cement-based composite we get higher flexural strength and load transfer capacity. Upon employing a free vibration test, Her and Lai (2013) explored the influence of CNTs (MW) on the structural dynamic response of epoxy-MWCNT PNCs. They observed that functionalization increases the interfacial bond among the CNTs and epoxy matrix, which diminishes its interfacial energy dissipation ability and leads to higher stiffness. Formica et al. (2010) implemented the Eshelby–MT approach by employing an equivalent continuum model for the study of vibration properties of CNTRC, which revealed that the lowest natural frequencies increase up to 500% for CNT–reinforced rubber composites without varying the mass density of the material in practice. Chen et al. (2011) analyzed the effect of tube curvature on the prepared PNCs' toughness. Their model predicted that peak bridging stress is increased in tube bending with decrement in average pull-out length.

1.2.5 RESEARCH WORKS ON AGGLOMERATION EFFECT OF CNTS ON PNCS CHARACTERISTICS

Numerous researchers in published literature examined the agglomeration effect of CNTs on the characteristics of CNT-reinforced PNCs either by using an analytical or experimental approach. Due to previous experimental investigations it is very well known that the CNTs agglomeration with higher or lower content considerably affects the various dynamic characteristics of the PNCs. Shi et al. (2004) explored the waviness and agglomeration of CNTs by employing the micromechanics methods. Eshelby's inclusion model is considered to analyse the influence of nanotubes agglomeration on stiffness. They also presented the connectivity among the morphology and effective properties of the CNT-reinforced PNCs, which are critical factors for enhancing and modifying their mechanical characteristics. Pourasghar et al. (2013) studied the effect of nanotubes agglomeration on the damping response of a CG-CNTR cylindrical panel employing the 3D theory of elasticity. The effective material properties are assessed employing the Eshelby–MT approach. To define the governing equations and boundaries conditions, an effective and precise numerical tool such as GDQM is employed. They found that the degree of agglomeration greatly influenced stiffness and vibration.

Yang et al. (2012) examined the thermo-mechanical response and morphology of CNT/shape memory polymer (SMP) composites. The primary aim of their investigation was to examine the effect of nanotubes agglomeration on the characteristics of the PNCs. They found that the agglomeration of CNTs dramatically affects the effective properties of the CNT/SMP composites while homogenous dispersion of CNT maximises the effective properties of the composite. Numerous experimental research on CNTRCs suggest that the excellent thermal and electrical properties of CNTs and hygro-thermal expansion coefficients of epoxies produce significant effects on the various properties of PNCs. Ghajar et al. (2017) examined the influence of agglomeration and dispersion on the viscoelastic characteristics of PNCs. They used the MT micromechanical method to examine effective properties of the PNCs–reinforced random-oriented CNTs. Their findings reported that by conducting more in-depth investigation of the agglomeration parameters as well as reducing the CNTs' agglomeration, properties of the nanocomposite can be easily enhanced.

1.3 SUMMARY

From the present chapter, the following conclusions from the literatures pertaining to the PMCs reinforced with CNTs have been drawn:

- Augmentation of the thermal, mechanical and vibrational characteristics of PNCs are significantly associated with the uniform exfoliation of CNTs. Moreover, the interfacial bond among the polymer matrix and CNTs is an important aspect and is responsible for the efficient load and heat transfer across the CNT–polymer interface.
- The thermal, mechanical, thermo-mechanical and vibrational characteristics of CNT–reinforced PMCs are significantly affected by numerous design parameters such as characteristics of the polymer matrix, loading of CNTs (V_{CNT}), aspect ratio, interfacial interactions with matrix, CNT diameter and chirality, orientation and dispersion of the CNTs in the matrix.
- It has been also found that the functionalization of CNTs increases the modulus, fracture resistance, strength, glass transition (Tg), thermal decomposition, melting temperatures and damping response of the CNT–reinforced PMCs. Moreover, functionalization of CNTs has dual benefits, such as improvement in the dispersion and good interfacial interactions among the matrix and CNTs, which are prerequisites for improvement in the PNCs' properties.
- It is found that the excessive employment of higher concentrated chemicals and acid during chemical functionalization may damage the structure of CNTs.
- To summarise, it is anticipated that PMC's ongoing growth opens up new possibilities for the next generation of cutting-edge aerospace systems and beyond.

REFERENCES

Ajayan, P.M., C. Colliex, and D. Trauth. "Aligned Carbon Nanotube Arrays Formed by Cutting a Polymer Resin-Nanotube Composite". *Science* 265, no. 5176 (1994): 1212–1214.

Bhattacharya, M. "Polymer Nanocomposites-A Comparison between Carbon Nanotubes, Graphene, and Clay as Nanofillers". *Materials* 9, no. 262 (2016): 1–35.

Biercuk, M.J., M.C. Llaguno, M. Radosavljevic, J.K. Hyun, A.T. Johnson, and J.E. Fischer. "Carbon Nanotube Composites for Thermal Management". *Applied Physics Letters* 80, no. 15 (2002): 2767–2769.

Bin-Jie, X., C. Zhuo-Ming, W. Xiang-Ji, and W. Xiao-Feng. "Study on the Thermal Behaviour of Polysulfonamide/ Multi-Wall Carbon Nanotube Composites". *Journal of Industrial Textiles* 42, no. 4 (2013): 434–445.

Chen, X., I.J. Beyerlein, and L.C. Brinson. "Bridged Crack Models for the Toughness of Composites Reinforced with Curved Nanotubes". *Journal of the Mechanics and Physics of Solids* 59, no. 9 (2011): 1938–1952. doi:10.1016/j.jmps.2010.12.012.

Chen, H., V. V. Ginzburg, J. Yang, Y. Yang, W. Liu, Y. Huang, L. Du, and B. Chen. "Thermal Conductivity of Polymer-Based Composites: Fundamentals and Applications". *Progress in Polymer Science* 59 (2016): 41–85. doi:10.1016/j.progpolymsci.2016.03.001.

Coleman, J.N., U. Khan, W.J. Blau, and Y.K. Gun'ko. "Small but Strong: A Review of the Mechanical Properties of Carbon Nanotube-Polymer Composites". *Carbon* 44, no. 9 (2006): 1624–1652.

Ferreira, F. V., B.R.C. Menezes, W. Franceschi, E. V. Ferreira, K. Lozano, L.S. Cividanes, A.R. Coutinho, and G.P. Thim. "Influence of Carbon Nanotube Concentration and Sonication Temperature on Mechanical Properties of HDPE/CNT Nanocomposites". *Fullerenes Nanotubes and Carbon Nanostructures* 25, no. 9 (2017): 531–539.

Fidelus, J.D., E. Wiesel, F.H. Gojny, K. Schulte, and H.D. Wagner. "Thermo-Mechanical Properties of Randomly Oriented Carbon/Epoxy Nanocomposites". *Composites Part A: Applied Science and Manufacturing* 36, no. 11 (2005): 1555–1561.

Formica, G., W. Lacarbonara, and R. Alessi. "Vibrations of Carbon Nanotube-Reinforced Composites". *Journal of Sound and Vibration* 329, no. 10 (2010): 1875–1889. doi:10.1016/j.jsv.2009.11.020.

Gaidukovs, S., V. Kampars, J. Bitenieks, I. Bochkov, G. Gaidukova, and U. Cabulis. "Thermo-Mechanical Properties of Polyurethane Modified with Graphite Oxide and Carbon Nanotube Particles". *Integrated Ferroelectrics* 173, no. 1 (2016): 1–11.

Ghajar, R., M.M. Shokrieh, and A.R. Shajari. "Investigation of Agglomeration and Dispersion of Reinforcement on the Viscoelastic Properties of CNT Reinforced Polymeric Composites". *AmirKabir Jounrnal of Science & Research Mechanical Engineering* 48, no. 4 (2017): 133–135.

Gojny, F.H., J. Nastalczyk, Z. Roslaniec, and K. Schulte. "Surface Modified Multi-Walled Carbon Nanotubes in CNT/Epoxy-Composites". *Chemical Physics Letters* 370, no. 5–6 (2003): 820–824.

Guo, P., X. Chen, X. Gao, H. Song, and H. Shen. "Fabrication and Mechanical Properties of Well-Dispersed Multiwalled Carbon Nanotubes/Epoxy Composites". *Composites Science and Technology* 67, no. 15–16 (2007): 3331–3337.

Her, S.C., and C.Y. Lai. "Dynamic Behavior of Nanocomposites Reinforced with Multi-Walled Carbon Nanotubes (MWCNTs)". *Materials* 6, no. 6 (2013): 2274–2284.

Jee, M.H., J.U. Choi, S.H. Park, Y.G. Jeong, and D.H. Baik. "Influences of Tensile Drawing on Structures, Mechanical, and Electrical Properties of Wet-Spun Multi-Walled Carbon Nanotube Composite Fiber". *Macromolecular Research* 20, no. 6 (2012): 650–657.

Jin, F.-L., and S.-J. Park. "A Review of the Preparation and Properties of Carbon Nanotubes-Reinforced Polymer Compositess". *Carbon Letters* 12, no. 2 (2011): 57–69.

Khalili, S.M.R., and A. Haghbin. "Investigation on Design Parameters of Single-Walled Carbon Nanotube Reinforced Nanocomposites under Impact Loads". *Composite Structures* 98 (2013): 253–260.

Kim, S.W., T. Kim, Y.S. Kim, H.S. Choi, H.J. Lim, S.J. Yang, and C.R. Park. "Surface Modifications for the Effective Dispersion of Carbon Nanotubes in Solvents and Polymers". *Carbon* 50, no. 1 (2012): 3–33.

Kwon, S.Y., I.M. Kwon, Y.G. Kim, S. Lee, and Y.S. Seo. "A Large Increase in the Thermal Conductivity of Carbon Nanotube/Polymer Composites Produced by Percolation Phenomena". *Carbon* 55 (2013): 285–290.

Li, W.W., W.M. Ji, Y. Liu, F. Xing, and Y.K. Liu. "Damping Property of a Cement-Based Material Containing Carbon Nanotube". *Journal of Nanomaterials* 2015 (2015): 371404. doi:10.1155/2015/371404.

Liao, K., and S. Li. "Interfacial Characteristics of a Carbon Nanotube-Polystyrene Composite System". *Applied Physics Letters* 79, no. 25 (2001): 4225–4227.

Ma, P.C., N.A. Siddiqui, G. Marom, and J.K. Kim. "Dispersion and Functionalization of Carbon Nanotubes for Polymer-Based Nanocomposites: A Review". *Composites Part A: Applied Science and Manufacturing* 41, no. 10 (2010): 1345–1367.

Manchado, M.A.L., L. Valentini, J. Biagiotti, and J.M. Kenny. "Thermal and Mechanical Properties of Single-Walled Carbon Nanotubes-Polypropylene Composites Prepared by Melt Processing". *Carbon* 43, no. 7 (2005): 1499–1505.

Mansour, G., D. Tzetzis, and K.D. Bouzakis. "A Nanomechanical Approach on the Measurement of the Elastic Properties of Epoxy Reinforced Carbon Nanotube Nanocomposites". *Tribology in Industry* 35, no. 3 (2013): 190–199.

Martone, A., C. Formicola, F. Piscitelli, M. Lavorgna, M. Zarrelli, V. Antonucci, and M. Giordano. "Thermo-Mechanical Characterization of Epoxy Nanocomposites with Different Carbon Nanotube Distributions Obtained by Solvent Aided and Direct Mixing". *Express Polymer Letters* 6, no. 7 (2012): 520–531.

Montazeri, A., J. Javadpour, A. Khavandi, A. Tcharkhtchi, and A. Mohajeri. "Mechanical Properties of Multi-Walled Carbon Nanotube/Epoxy Composites". *Materials and Design* 31, no. 9 (2010): 4202–4208.

Montazeri, A., and N. Montazeri. "Viscoelastic and Mechanical Properties of Multi Walled Carbon Nanotube/Epoxy Composites with Different Nanotube Content". *Materials and Design* 32, no. 4 (2011): 2301–2307.

Montazeri, A., N. Montazeri, and S. Farzaneh. "Thermo-Mechanical Properties of Multi-Walled Carbon Nanotube (Mwcnt)/Epoxy Composites". *International Journal of Polymer Analysis and Characterization* 16, no. 3 (2011): 199–210.

Pande, S., R.B. Mathur, B.P. Singh, and T.L. Dhami. "Synthesis and Characterization of Multiwalled Carbon Nanotubes-polymethyl Methacrylate Composites Prepared by in Situ Polymerization Method". *Polymer Composites* 30, no. 9 (2009): 1312–1317.

A. Pourasghar, M.H. Yas, S. Kamarian. "Local Aggregation Effect of CNT on the Vibrational Behavior of Four-Parameter Continuous Grading Nanotube-Reinforced Cylindrical Panels". *Polymer Composites* (2013): 708–721.

Qian, D., E.C. Dickey, R. Andrews, and T. Rantell. "Load Transfer and Deformation Mechanisms in Carbon Nanotube-Polystyrene Composites". *Applied Physics Letters* 76, no. 20 (2000): 2868–2870.

Rahmat, M., and P. Hubert. "Carbon Nanotube-Polymer Interactions in Nanocomposites: A Review". *Composites Science and Technology* 72, no. 1 (2011): 72–84.

Safadi, B., R. Andrews, and E.A. Grulke. "Multiwalled Carbon Nanotube Polymer Composites: Synthesis and Characterization of Thin Films". *Journal of Applied Polymer Science* 84, no. 14 (2002): 2660–2669.

Sharma, S., R. Chandra, P. Kumar, and N. Kumar. "Thermo-Mechanical Characterization of Multi-Walled Carbon Nanotube Reinforced Polycarbonate Composites: A Molecular Dynamics Approach". *Comptes Rendus - Mecanique* 343, no. 5–6 (2015): 371–396.

Shi, D.-L., X.-Q. Feng, Y.Y. Huang, K.-C. Hwang, and H. Gao. "The Effect of Nanotube Waviness and Agglomeration on the Elastic Property of Carbon Nanotube-Reinforced Composites". *Journal of Engineering Materials and Technology* 126, no. 3 (2004): 250.

Soni, S.K., B. Thomas, and V.R. Kar. "A Comprehensive Review on CNTs and CNT-Reinforced Composites: Syntheses, Characteristics and Applications". *Materials Today Communications* 25, no. August 2020: 101546.

Spitalsky, Z., L. Matějka, M. Šlouf, E.N. Konyushenko, J. Kovářová, J. Zemek, and J. Kotek "Modification of Carbon Nanotubes and Its Effect on Properties of Carbon Nanotube/ Epoxy Nanocomposites". *Polymer Composites* 30, no. 10 (2009): 1378–1387.

Spitalsky, Z., D. Tasis, K. Papagelis, and C. Galiotis. "Progress in Polymer Science Carbon Nanotube – Polymer Composites: Chemistry, Processing, Mechanical and Electrical Properties". *Progress in Polymer Science* 35, no. 3 (2010): 357–401.

Velasco-Santos, C., A.L. Martınez-Hernandez, F. Fisher, R. Ruoff, and V.M. Castano. "Dynamical-Mechanical and Thermal Analysis of Carbon Nanotube-Methyl-Ethyl Methacrylate Nanocomposites". *Journal of Physics D: Applied Physics* 36 (2003): 1423–1428.

Wang, S., R. Liang, B. Wang, and C. Zhang. "Dispersion and Thermal Conductivity of Carbon Nanotube Composites". *Carbon* 47, no. 1 (2009): 53–57.

Wernik, J.M., and S.A. Meguid. "Recent Developments in Multifunctional Nanocomposites Using Carbon Nanotubes". *Applied Mechanics Reviews* 63, no. 5 (2010): 050801.

Xiong, J., Z. Zheng, X. Qin, M. Li, H. Li, and X. Wang. "The Thermal and Mechanical Properties of a Polyurethane/Multi-Walled Carbon Nanotube Composite". *Carbon* 44, no. 13 (2006): 2701–2707.

Yang, Q.S., X.Q. He, X. Liu, F.F. Leng, and Y.W. Mai. "The Effective Properties and Local Aggregation Effect of CNT/SMP Composites". *Composites Part B: Engineering* 43, no. 1 (2012): 33–38.

Yang, Z., D. Xu, J. Liu, J. Liu, L. Li, L. Zhang, and J. Lv. "Fabrication and Characterization of Poly(Vinyl Alcohol)/Carbon Nanotube Melt-Spinning Composites Fiber". *Progress in Natural Science: Materials International* 25, no. 5 (2015): 437–444.

Yu, A., M.E. Itkis, E. Bekyarova, and R.C. Haddon. "Effect of Single-Walled Carbon Nanotube Purity on the Thermal Conductivity of Carbon Nanotube-Based Composites". *Applied Physics Letters* 89, no. 13 (2006): 2004–2007.

2 A Review on Natural Fibre Reinforced Composite Under UV Concentration for Aircraft Application

Syed Noor Hussain bin Mohamed Hasanuthin

Universiti Tun Hussein Onn Malaysia (UTHM), Parit Raja, Malaysia

A Hamdan

Universiti Tun Hussein Onn Malaysia (UTHM), Parit Raja, Malaysia

Universiti Malaysia Perlis, Kangar, Malaysia

A R Irfan

Universiti Malaysia Perlis (UniMAP), Arau, Malaysia

Universiti Malaysia Perlis, Kangar, Malaysia

Mohammad Sukri Mustapa and Mohamad Norani Mansor

Universiti Tun Hussein Onn Malaysia (UTHM), Parit Raja, Malaysia

CONTENTS

DOI: 10.1201/9781003200994-2

2.1 INTRODUCTION

Aircraft are a magnificent accomplishment of engineering. Humankind had figured out how to get metal weighing more than 100,000 pounds to fly flatly, without any mistakes. Aircraft components have advanced over the last decade to improve performance, structural integrity, reliability and speed in flights (Katnam, Da Silva, and Young 2013). There are a few aircraft systems such as fuselage, wing, empennage, radome and the aircraft's interior (Katnam, Da Silva, and Young 2013). The fuselage is the biggest surface area in an aircraft. Most of the fuselage components already employ composite materials. The radome also employs composite as its material.

The fuselage is a long hollow tube, and it is one of the main elements of the aeroplane, often referred to as the aircraft body that carries passengers and luggage. The cockpit is also included as part of the fuselage. The fuselage has various types, but the function is the same – to connect the aircraft's main components (Katnam, Da Silva, and Young 2013). Aircraft fuselages are made up of thin sheets of material solidified with loads of longitudinal stringers along with cross-frame and dowel racks. They usually carry bending, shear forces, torsional loads produced in the chains, and the skin axial pressures and the shear stress in the skin; the tensions for cord strength usually are ignored (Megson 2013). The shape of the fuselage usually will depend on aircraft duty. A supersonic fighter plane has a very slim, streamlined fuselage to reduce the drag involved in high-speed operations, whereas for commercial aircraft, the fuselage is usually wider to carry the maximum quantity of passengers and luggage. In airliners, pilots will be seated in a cockpit located at the front of the fuselage. The rear of the fuselage is where the passengers and luggage are located, and the petrol for the aircraft will be stored in the wings. For fighter jets, the cockpit is usually located on the tip of the fuselage, armour is placed at the wings, and the petrol and the engine are located at the tail of the fuselage. The weight of the aircraft is divided through the whole length of the aircraft. The fuselage that carries passengers and luggage counts as the essential part of the aircraft's mass. The aircraft's centre of gravity is where the average weight of the entire aircraft lies, and it is located inside the fuselage. Due to the torque produced by the elevator, rudder and ailerons, the plane rotates around the centre of gravity during flight. In order to withstand the strength produced by these torques, a fuselage needs to be equipped with adequate strength (Katnam, Da Silva, and Young 2013).

The radome is a term that refers to a radar and dome hybrid. The radar on commercial and military aircraft, as well as some helicopters, is protected by an aerodynamically shaped dome. The radome is located in the aircraft's nose. On military aircraft, such as jets, the length of the aircraft radome can be extended. The radome's conical configuration aids the aircraft's aerodynamics and is made of composite materials, most commonly glass fibre. A forward-looking radar system for weather detection is installed in the radome of commercial aircraft. The radome protects the weather radar on the nose of the aircraft. Weather radars operate at particular frequencies, which the radome assembly must efficiently transmit around to enable the tool to operate correctly. Radomes are vulnerable to many types of damages, including rain, bird strikes and damage caused by ground infrastructures such as hangar doors and repair equipment (Haris et al. 2011). Any kind of damage to the radome will make it susceptible to moisture flow into the fibreglass structures, which may prevent the unique frequency from hitting the weather radar. A transmission performance test calculates how effectively the weather radar transmits and absorbs electromagnetic activity (Botelho, Nohara, and Rezende 2015).

These are two examples of exterior parts in aircraft that use composite as the material. The interior parts are not exposed to high load and extreme weather; hence, structural integrity and strength play a minor role.

2.2 INTERIOR OF AIRCRAFT

The interior of an aircraft plays a vital role in providing a conducive environment for the cabin crew. Plastic products are used for a wide range of aircraft interior parts and assemblies. The aircraft cabin is a crucial part of the aircraft's interior, containing various specialised plastic materials. Interior materials include seats, composite tables, in-flight entertainment systems, curtains, windscreens, bins and lighting systems. Skilled Plastics provides products engineered to meet the specifications of the Federal Aviation Authority for a blaze, heat escape and smoke generation. Skilled Plastics has received ISO 9001, ISO 14001 and AS9100D certifications and contract awards from Boeing, Lockheed and others (Dos Santos et al. 2016; Vink et al. 2012).

The tray table in the aircraft is the table that is attached to the passengers' seats. This tray table is often found in the economy class in the aircraft. The tray table may not seem to be necessary, but it has its function. The tray table will be helpful for passengers to enjoy their meal conveniently during their flight. It will also be helpful for passengers who want to do their work or assignment using their laptops, reading books or writing. The material chosen in the manufacturing of the tray table is essential as it will also contribute to the aircraft's total weight. One tray table might not have much weight, but the there will be around 200 to 300 passengers' seats in economy class. Therefore, the total weight will affect the weight of the aircraft. Besides that, the material also needs to have high strength. After the tray table is fixed during manufacturing, the same table will be used until the service period of the aircraft ends. Composite materials are used in the tray table. However, the material used is a synthetic material, which has sustainability issues. Nevertheless, an upcoming study shows that a newly developed hybrid composite will be used to manufacture the tray table and is expected to have lower maintenance concerns than the current material (Nor, Sultan, and Hamdan 2017; Gomez-Campos et al. 2021).

Another airline part is the cabin windows. The material is a lightweight composite called "stretched acrylic". Stretched acrylic is made by stretching the as-cast acrylic base content. It has a higher crack tolerance [hairline cracks], a smaller crack distribution and better impact resistance. Compared to the cabin windows, the flight deck windshields are made of acrylic glass—the outer layer of glass fused to acrylic. There is a coating between them, composed of urethane (Katnam, Da Silva, and Young 2013; Mike Arnot 2019).

The discussion on the tray table and cabin windows exhibits the importance of material selection to ensure sustainability of the parts. The cabin windows are interior parts exposed directly to ultraviolet radiation and tend to deteriorate from their original characteristic. The understanding of the composite is explained in the next paragraph.

2.3   SYNTHETIC COMPOSITE MATERIAL

Many synthetic materials are the products of chemical reactions in which the original compounds' atomic structures are rearranged to create a new material. Synthetic fibres are versatile. They can be used to create clothes and other things. Plastics are synthetic materials that are very useful in our everyday life. Plastics can be moulded when they are soft into various forms that will harden later on. Humans have learned how to make plastics from a range of ingredients, but today the most common plastics are manufactured from fossil fuels such as petroleum (Supian et al. 2018). Polymers are solid and versatile. They can be conveniently changed to hold various shapes and colours. Examples of polymers include polyester, nylon and acrylic. Many common artefacts are constructed from synthetic polymers. You may be acquainted with some of them. Pipes made of polyvinyl chloride (PVC) are used for plumbing. House siding, rainwear and garden hoses are made of polyvinyl chloride. Most types of plastic ropes are constructed from polypropylene. The main issues with synthetic composite materials are the hazardous manufacturing process and waste management problem (Ismail et al. 2019; Farinha, de Brito, and Veiga 2019). Therefore, most researchers are looking to reduce the application of synthetic fibres or even eliminate the usage of synthetic fibres (Nair and Laurencin 2007; Jauhari, Mishra, and Thakur 2015).

2.4   NATURAL FIBRE COMPOSITE MATERIAL

2.4.1   KENAF FIBRES

Natural fibre such as kenaf, flax, pineapple leaf and bamboo can reduce dependence on the synthetic fibre composite as a structural material (Venkateshwar Reddy et al. 2020). Hibiscus Cannabinus, L., or kenaf fibre, is an annual cotton and okra-related fibre crop. It is native to Africa, where it has been cultivated for at least 4,000 years for use in ropes and animal use. The kenaf stem is divided into two sections: the bark or bast, which contains relatively long fibres, and the heart, which contains short fibres. The core and bast make up respectively 65% and 35% of the overall weight of the stem. Considerable research has been carried out which has shown that kenaf is an environmentally sustainable plant with the ability to substitute for products such as synthetic polymers, glass fibres and wood (Mohammed et al. 2015).

2.4.2   PINEAPPLE LEAF FIBRE

Pineapple leaf fibre is a significant waste material in Malaysia that has yet to be investigated (Venkateshwar Reddy et al. 2020). From a socio-economic standpoint, pineapple leaf fibre can be a modern source of raw materials for the industry and eventually replace costly and non-renewable conventional fibres. Pineapple leaf fibre, on the other hand, has a high basic strength and hardness, as well as being hydrophilic due to its high cellulose content. Pineapple leafe fibre has greater mechanical strength than jute as used in the manufacture of fine yarn. Furthermore, it has a high flexural and torsional rigidity. Therefore it can be used as an excellent replacement raw material to strengthen composite matrixes because of its unique properties (Asim et al. 2015).

2.4.3 BAMBOO FIBRES

Bamboo (Bambusoideae subfamily) is a Poaceae family subfamily of tall tree-like grasses with over 115 genera and 1,400 species. Bamboos are usually fast-growing perennials, in some cases growing up to 30 cm per day. Culms are woody ringed roots that are hollow between rings (nodes) and appear in clusters of thick rhizomes (underground stem). Bamboo culms can grow from 10 to 15 cm in the smallest species to more than 40 metres in the tallest. Though narrow leaves appear directly from the stem rings on young culms, mature culms often sprout horizontal leaf-bearing branches. Bamboo jointed stems may be used for a variety of purposes; the larger stems are used to make planks for houses and rafts, while both large and small stems are used to make scaffolding for construction sites (Venkateshwar Reddy et al. 2020). Interestingly, the bamboo fibre has potential to reinforce composite material (Hojo et al. 2014)

2.5 COMPONENT COMPOSITION IN NATURAL FIBRE

The three essential components of natural plant fibre are cellulose, hemicellulose and lignin. The three component compositions' content can influence the mechanical properties of the fibres. Some studies treated the fibres in order to improve the component composition contents. Several researchers used an alkaline solution to increase the tensile strength of the fibres, in which the treatment partly extracts all hemicellulose and lignin, which serves as a cellulose binder. As a consequence, the lower the lignin content, the higher the cellulose content, and hence the higher the tensile strength of the fibre (Hassan et al. 2018; Guo, Sun, and Satyavolu 2019).

As an example for the kenaf fibre, cellulose was the most dominant content in the natural fibre, however, some researchers treated the fibres with an alkaline treatment in order to improve the composition content. The comparison of component composition between untreated and treated kenaf fibre was summarised in Table 2.1.

The cellulose in pineapple leaf is the dominant content in the natural fibre. Generally the content of cellulose, hemicellulose and lignin is about 70 ± 10 percentage, 10 ± 10 percentage, and 8 ± 8 percentage respectively. Pineapple leaf fibre is a more compatible natural fibre resource with a good component composition as compared to others. The treated pineapple leaf fibre possesses higher cellulose content than the untreated fibre, where the tensile strength is higher (Mahardika et al. 2018; Fareez et al. 2018). The component comparison between the untreated and treated pineapple leaf is tabulated in Table 2.2. Therefore, it is crucial to understand the characteristic to suit the structural requirements.

TABLE 2.1

Component Composition of Untreated and Treated Kenaf Fibre

Treatment	Cellulose (%)	Hemicellulose (%)	Lignin (%)	References
Untreated	44.00–57.00	21.00	15.00–19.00	(Alavudeen et al. 2015)
Untreated	60.46–66.24	12.60–19.91	14.67–19.24	(Wang et al. 2016)
Untreated	45.00–57.00	21.50	8.00–13.00	(Akil et al. 2011)
Untreated	72.00	20.30	9.00	(Yusuff, Sarifuddin, and Ali 2020)
Untreated	70.00	19.00	3.00	(Millogo et al. 2015)
Treated	74.10	12.20	6.30	(Hassan et al. 2018)
Treated	79.30	9.69	7.22	(Guo, Sun, and Satyavolu 2019)

TABLE 2.2

Component Composition of Untreated and Treated Pineapple Leaf Fibre

Treatment	Cellulose (%)	Hemicellulose (%)	Lignin (%)	References
Untreated	80.00–81.00	16.00–19.00	4.60–12.00	(Jose, Salim, and Ammayappan 2016)
Untreated	66.20	19.50	4.20	(Daud, Awang, et al. 2014a)
Untreated	62.50	13.90	15.9	(Mahardika et al. 2018)
Untreated	70.00–82.00	–	5.00–12.00	(Mishra et al. 2004)
Untreated	66.20	19.50	4.28	(Daud, Hatta, et al. 2014b)
Treated	81.30	2.90	1.50	(Mahardika et al. 2018)
Treated	85.53 ± 2.30	0.30 ± 0.90	0.40 ± 0.30	(Fareez et al. 2018)

2.6   ULTRAVIOLET (UV) TEST

Ultraviolet (UV) radiation is a kind of energy emitted by the sun. UV radiation has a shorter wavelength than visible light, but it is harmless to the naked eye. The human skin, on the other hand, can detect UV radiation. The sun emits a wide range of ultraviolet radiation, which is classified as UV-A, UV-B and UV-C. The ozone layer will completely filter the UV-C before entering the earth surface as this ray is very dangerous to the human body. The UV-B could enter the earth surface and cause sunburn if human skin were exposed to this ray. Additionally, when human skin is exposed to UV-B for a long time, it can cause skin disease such as skin cancer and damage the human body's DNA. However, only a small percentage of UV-B enters the earth surface as 95% of UV-B is filtered by our ozone layer ("Ultraviolet Waves," n.d.). As for UV-A, it is the least dangerous of UV radiation. UV-A has a longer wavelength compared to UV-B and UV-C, thus less effect on human skin. It only causes skin ageing if we're exposed to UV-A (Anna Chien and Jacobe, n.d.).

Generally, several types of research were conducted to study the effect of UV radiation on composite materials. The material's qualities such as tensile strength and glass transitiondropped after being exposed to UV radiation. Table 2.3 summarises the findings conducted by several researchers on the UV effect on the composite.

TABLE 2.3
The Effect of UV Radiation on the Composite Materials

Material	UV expose duration	Performance	References
Polyurethane composite materials	50 hours, 100 hours	The mechanical properties performance is decreased	(Gorbunov, Berdnikova, and Poluboyarov 2019)
Glass Reinforced Polymer (GRP) composites	Exposure to Ultraviolet radiation. Increased temperature for about 1000 h	Proposed model of synergistic ageing under UV and water condensation	(Lu et al. 2016)
Glass fiber/epoxy composites	15 days, 35 days, 45 days	The degradation effect is clearly presented	(Lohani et al. 2021)
Styrene-based shape memory polymers	UV wavelength in the range of 325–400 nm	The mechanical properties and the glass transition temperature (Tg) drop considerably drop	(Al Azzawi, Epaarachchi, and Leng 2018)

Most of the research conducted is on synthetic fibres. There is still a need for an investigation on natural fibre, especially under the effects of UV radiation. As the window shade is one of the interior components in the aircraft that is most exposed to the sunlight during flight, more effort is required to investigate the application of natural fibre in an aircraft application.

2.7 CONCLUSION

The safety aspects of the aircraft are always the most crucial aspect of the aviation industry. The material used in the aircraft components needs to be studied and upgraded over time to increase the aircraft's safety. Additionally, research also needs to be conducted to discover materials that can reduce the cost of aircraft manufacturing as much as possible but at the same time have high strength and durability.

REFERENCES

Akil, H M, M F Omar, A A M Mazuki, S Safiee, Z A M Ishak, and A Abu Bakar. 2011. "Kenaf Fiber Reinforced Composites: A Review." *Materials and Design* 32: 4107–4121.

Alavudeen, A., N. Rajini, S. Karthikeyan, M. Thiruchitrambalam, and N. Venkateshwaren. 2015. "Mechanical Properties of Banana/Kenaf Fiber-Reinforced Hybrid Polyester Composites: Effect of Woven Fabric and Random Orientation." *Materials and Design* 66 (PA): 246–257. doi:10.1016/j.matdes.2014.10.067.

Anna Chien, MD, and M.D. Heidi Jacobe. n.d. "UV Radiation & Your Skin, The Facts, The Risks, How They Affect You." 2019. https://www.skincancer.org/risk-factors/uv-radiation/.

Asim, M, K. Abdan, M. Jawaid, M. Nasir, Z. Dashtizadeh, M. R. Ishak, and M. E. Hoque. 2015. "A Review on Pineapple Leaves Fibre and Its Composites" 2015.

Azzawi, W., J. A. Epaarachchi, and Jinsong Leng. 2018. "Investigation of Ultraviolet Radiation Effects on Thermomechanical Properties and Shape Memory Behaviour of Styrene-Based Shape Memory Polymers and Its Composite." *Composites Science and Technology*. doi:10.1016/j.compscitech.2018.07.001.

Botelho, E. C., E. L. Nohara, and M. C. Rezende. 2015. "Lightweight Structural Composites with Electromagnetic Applications." *Multifunctionality of Polymer Composites: Challenges and New Solutions.* doi:10.1016/B978-0-323-26434-1.00012-X.

Daud, Z., H. Awang, A. S. M. Kassim, M. Z. M. Hatta, and A. M. Aripin. 2014a. "Comparison of Pineapple Leaf and Cassava Peel by Chemical Properties and Morphology Characterization." In *Advanced Materials Research*, 974:384–388. Trans Tech Publications Ltd. doi:10.4028/www.scientific.net/AMR.974.384.

Daud, Z., M. Z. M. Hatta, A. S. M. Kassim, H. Awang, and A. M. Aripin. 2014b. "Exploring of Agro Waste (Pineapple Leaf, Corn Stalk, and Napier Grass) by Chemical Composition and Morphological Study." *BioResources.* https://doi.org/10.15376/biores.9.1.872-880.

Farinha, C. B., J. de Brito, and R. Veiga. 2019. "Assessment of Glass Fibre Reinforced Polymer Waste Reuse as Filler in Mortars." *Journal of Cleaner Production* 210: 1579–1594. doi:10.1016/j.jclepro.2018.11.080.

Gomez-Campos, A., C. Vialle, A. Rouilly, L. Hamelin, A. Rogeon, D. Hardy, and C. Sablayrolles. 2021. "Natural Fibre Polymer Composites - A Game Changer for the Aviation Sector?" *Journal of Cleaner Production.* doi:10.1016/j.jclepro.2020.124986.

Gorbunov, F. K., L. K. Berdnikova, and V. A. Poluboyarov. 2019. "The Influence of Carbide Modifiers on Performance of Polyurethane Exposed to High Temperature and UV Radiation." *Materials Today: Proceedings.* doi:10.1016/j.matpr.2020.01.300.

Fareez, I.M., Ibrahim, N.A., Yaacob, W.M.H.W., Razali, N.A.M., Jasni, A.H. and Aziz, F.A. 2018. "Characteristics of Cellulose Extracted from Josapine Pineapple Leaf Fibre after Alkali Treatment Followed by Extensive Bleaching." *Cellulose* 25 (8): 4407–4421. doi:10.1007/s10570-018-1878-0.

Guo, A., Z. Sun, and J. Satyavolu. 2019. "Impact of Chemical Treatment on the Physiochemical and Mechanical Properties of Kenaf Fibers." *Industrial Crops and Products* 141 (December). doi:10.1016/j.indcrop.2019.111726.

Haris, M. Y., D. Laila, E. S. Zainudin, F. Mustapha, R. Zahari, and Z. Halim. 2011. "Preliminary Review of Biocomposites Materials for Aircraft Radome Application." *Key Engineering Materials.* doi:10.4028/www.scientific.net/KEM.471-472.563.

Hassan, Aziz, M. R. Mohd Isa, Z. A. Mohd Ishak, N. A. Ishak, N. A. Rahman, and F. M. Salleh. 2018. "Characterization of Sodium Hydroxide-Treated Kenaf Fibres for Biodegradable Composite Application." *High Performance Polymers* 30 (8): 890–899. doi:10.1177/0954008318784997.

Hojo, T., X. U. Zhilan, Y. Yang, and H. Hamada. 2014. "Tensile Properties of Bamboo, Jute and Kenaf Mat-Reinforced Composite." *Energy Procedia* 56 (C): 72–79. doi:10.1016/j.egypro.2014.07.133.

Ismail, M F B, M T H Sultan, A Hamdan Ariffin, A U M Shah, M Jawaid, and S N A Safri. 2019. "Kenaf/Glass Hybrid Composites." *International Journal of Recent Technology and Engineering* 8 (1): 456–461. https://www.scopus.com/inward/record.uri?eid=2-s2.0-85068456803&partnerID=40&md5=3fef77f0cf76e5aed7f61fdba9360eb5.

Jauhari, N., Mishra, R. and Thakur, H. 2015. "Natural Fibre Reinforced Composite Laminates - A Review." In *Materials Today: Proceedings* 2(4–5) doi:10.1016/j.matpr.2015.07.304.

Jose, S., Salim, R. and Ammayappan, L. 2016. "An Overview on Production, Properties, and Value Addition of Pineapple Leaf Fibers (PALF)." *Journal of Natural Fibers.* Taylor and Francis Inc. doi:10.1080/15440478.2015.1029194.

Katnam, K. B., L. F.M. Da Silva, and T. M. Young. 2013. "Bonded Repair of Composite Aircraft Structures: A Review of Scientific Challenges and Opportunities." *Progress in Aerospace Sciences* 61: 26–42. doi:10.1016/j.paerosci.2013.03.003.

Lohani, S., Prusty, R.K. and Ray, B.C. 2021. "Effect of Ultraviolet Radiations on Interlaminar Shear Strength and Thermal Properties of Glass Fiber/Epoxy Composites." *Materials Today: Proceedings.* doi:10.1016/j.matpr.2020.12.028.

Lu, T., Solis-Ramos, E., Yi, Y.B. and Kumosa, M. 2016. "Synergistic Environmental Degradation of Glass Reinforced Polymer Composites." *Polymer Degradation and Stability*. doi:10.1016/j.polymdegradstab.2016.06.025.

Mahardika, M., Abral, H., Kasim, A., Arief, S. and Asrofi, M. 2018. "Production of Nanocellulose from Pineapple Leaf Fibers via High-Shear Homogenization and Ultrasonication." *Fibers* 6 (2): 1–12. doi:10.3390/fib6020028.

Megson, T. H.G. 2013. Aircraft Structures for Engineering Students. *Aircraft Structures for Engineering Students*. doi:10.1016/C2009-0-61214-9.

Mike Arnot. 2019. "What Are Airplane Windows Made of? (And What Is That Little Hole?)." 2019. https://thepointsguy.com/news/what-are-airplane-windows-made-of/#:~:

Millogo, Y., Aubert, J.E., Hamard, E. and Morel, J.C. 2015. "How Properties of Kenaf Fibers from Burkina Faso Contribute to the Reinforcement of Earth Blocks." *Materials* 8 (5): 2332–2345. doi:10.3390/ma8052332.

Mishra, S., Mohanty, A.K., Drzal, L.T., Misra, M. and Hinrichsen, G. 2004. "A Review on Pineapple Leaf Fibers, Sisal Fibers and Their Biocomposites." *Macromolecular Materials and Engineering*. doi:10.1002/mame.200400132.

Mohammed, L., Ansari, M.N., Pua, G., Jawaid, M. and Islam, M.S. 2015. "A Review on Natural Fiber Reinforced Polymer Composite and Its Applications." *International Journal of Polymer Science*. doi:10.1155/2015/243947.

Nair, L. S., and C. T. Laurencin. 2007. "Biodegradable Polymers as Biomaterials." *Progress in Polymer Science (Oxford)*. doi:10.1016/j.progpolymsci.2007.05.017.

Nor, A F M, M T H Sultan, and A Hamdan. 2017. "Design and Development of a Food Tray Table for Commercial Aircraft Using Hybrid Composites." In *Proceedings of Mechanical Engineering Research Day 2017*.

Santos, C. V. D., D. R. Leiva, F. R. Costa, and J. A. R. Gregolin. 2016. "Materials Selection for Sustainable Executive Aircraft Interiors." *Materials Research*. doi:10.1590/1980-5373-MR-2015-0290.

Supian, A. B.M., S. M. Sapuan, M. Y.M. Zuhri, E. S. Zainudin, and H. H. Ya. 2018. "Hybrid Reinforced Thermoset Polymer Composite in Energy Absorption Tube Application: A Review." *Defence Technology*. doi:10.1016/j.dt.2018.04.004.

"Ultraviolet Waves." n.d. https://science.nasa.gov/ems/10_ultravioletwaves.

Venkateshwar Reddy, P., R.V. Saikumar Reddy, J. Lakshmana Rao, D. Mohana Krishnudu, and P. Rajendra Prasad. 2020. "An Overview on Natural Fiber Reinforced Composites for Structural and Non-Structural Applications." *Materials Today: Proceedings*. doi:10.1016/j.matpr.2020.10.523.

Vink, P., C. Bazley, I. Kamp, and M. Blok. 2012. "Possibilities to Improve the Aircraft Interior Comfort Experience." *Applied Ergonomics*. doi:10.1016/j.apergo.2011.06.011.

Wang, C., S. Bai, X. Yue, B. Long, and L. Choo-Smith. 2016. "Relationship between Chemical Composition, Crystallinity, Orientation and Tensile Strength of Kenaf Fiber." *Fibers and Polymers* 17 (11): 1757–1764. doi:10.1007/s12221-016-6703-5.

Yusuff, I., N. Sarifuddin, and A. M. Ali. 2020. "A Review on Kenaf Fiber Hybrid Composites: Mechanical Properties, Potentials, and Challenges in Engineering Applications." *Progress in Rubber, Plastics and Recycling Technology*, September, 147776062095343. doi:10.1177/1477760620953438.

3 A Review on Advanced Polymer Composites Used in Aerospace Application

Soundhar A

Sri Venkateswara College of Engineering (SVCE), Chennai, India

Lakshmi Narayanan V and Anita Jessie

Vellore institute of Technology, Vellore, India

Natesh M

V.S.B. Engineering College, Karur, India

CONTENTS

3.1 INTRODUCTION

Aerospace industries design and manufacture spacecraft and airplanes. In recent decades, the development of these industries has been remarkable. Structural design and the materials used are the major parameters considered for improving the performance of aerospace systems (Irving and Soutis 2019). Materials play a vital part in the lifecycle of an airplane system right from design, manufacturing, maintenance and disposal (till the final stage). The materials used in the aircraft are subjected to extreme loads from take-off to the landing stage. Some important requirements for aerospace materials (Ghori et al. 2018) are given below.

DOI: 10.1201/9781003200994-3

- High strength
- Light weight
- High damage tolerance
- Good fracture resistance

The aircraft material is generally used in the engine, rocket motor casting, antenna dishes, centre wind box, landing gear doors, engine cowls, horizontal and vertical stabilisers, floor beams, flaps, etc. A few safety-critical structures such as fuselage, wings, and landing gears, etc. are among the components used in the aircraft. Years ago, most of the structural components are were built using metals. However, in recent years, innovation in material sciences has grown to a larger extent. Thus, currently, metal structures are replaced by polymer composite materials (Joshi and Chatterjee 2016). Figure 3.1 shows the percentage of polymer composites utilised in specific aircraft (A300, A310-200, A320, A340-300, A340-600, A380, A400M and A350-900XWB) from 1970 to the present date.

Moreover, the polymer composite satisfies most of the necessary requirements for aircraft materials. Some of polymer composite's advantages (Gupta 2007) are listed below:

- Improves fatigue and resolves the corrosion issue associated with metals.
- Effective damage tolerance, thus increasing accidental survivability.
- High impact resistance decreases the damage to crucial engine pylons (the region associated with the fuel lines and engine controls).
- Mechanical properties of the composites can be easily improved by lay-up design and material orientation.

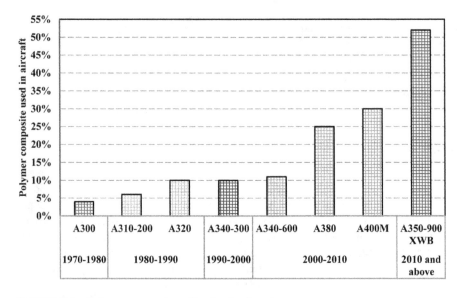

FIGURE 3.1 Polymer composite utilization in aircraft.

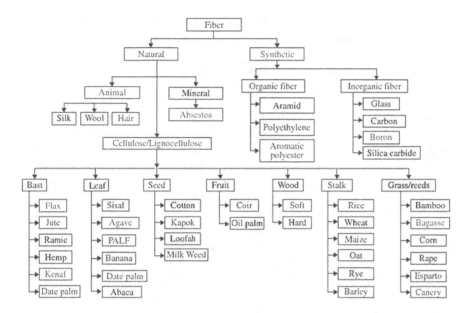

FIGURE 3.2 Different Types of fibres. (Saba et al. 2015)

- The complex composite parts can be easily assembled through a rotational molding process and automated lay-up machineries.
- Polymer composites have good chemical features, low cost and are lightweight.

Figure 3.2 shows the branches of fibres that are mostly used in day-to-day life (Saba et al. 2015).

With their huge advantages, polymer composites play a vital role in aerospace components and structures. Currently, plenty of research is being carried out in this area. The different types of polymer composites associated with the inner and outer parts of the aircraft and aerospace structures are discussed in detail in this chapter.

3.2 NATURAL FIBRE POLYMER COMPOSITES FOR AEROSPACE APPLICATION

Drastic expansion of industrial sectors resulted in an increase in carbon dioxide in the atmosphere, which resulted in a change in climatic conditions. On recognizing the effect of the climatic change, the demand for natural polymer composites has increased exponentially. So, most industries and researchers currently are showing interest to utilise renewable resources (eco-friendly materials) to develop natural polymer composites (Asim et al. 2018). The most preferred natural fibres to develop natural polymer composites are oil palm, jute, sisal, PALF, flax and kenaf (Ku et al. 2011). Incorporating a natural fibre that has a similar higher specific strength and specific modulus is increasing exponentially, providing plenty of advantages. Moreover, with cheaper sources and superior characteristics, natural fibres provide the desired modulus and strength at a low cost (Huda et al. 2006).

For centuries, natural fibres performance held a significant role in the fabrication of natural composite. Egyptians blended Nil's mud and wheat straws as a reinforcement to improve the strength of bricks and keep the house cooler during summer and warmer during the cold season. Natural fibres are degradable, renewable and sustainable, which lead to the invention of Natural fibre polymer composite (NFPC) (Mansor et al. 2019). An investigation on ramie-fibre–reinforced polymer composites in aircraft wing box components application was conducted by Boegler et al. (2015). A substantial reduction in weight (up to 14%) was observed when compared with aluminum alloy (7000 series). Aircraft interiors are made of hazardous and non-renewable materials such as phenolic resins, glass, carbon fibres and so on. These materials can be replaced by NFPC, which are eco-friendly materials made of flax, natural fibres and reinforced natural polymer composites. Eloy et al. (2015) discovered the possibility of using cotton fibre and bio-based polyurethane (BPU) (derived from castor bean oil) composite for aerospace interior components. The developed composite was compared with other polymer composites (glass-fibre–reinforced BPU, cotton-fibre–reinforced epoxy and glass -fibre–reinforced epoxy). Results revealed that BPU exhibited relatively lower tensile strength than its counterparts. However, the tensile strength was very close to cotton-fibre–epoxy composites. Therefore, bio-based composites have the potential to replace synthetic composites in interiors to create green aircraft.

Red (2012) reported that the requirement of composites in aircraft construction will double by 2022. This is due to the properties of polymer composites such as low cost, high strength, low weight, low flammability, durability and high manufacturing production capability. Manan et al. (2016) investigated NFPC–based honeycomb structures for aircraft structures subjected to high loads. The NFPC–based honeycomb structure, developed based on a vacuum-infusion technique, was exposed to various tests and the results show that the developed structure had superior mechanical properties, energy dissipation characteristics, better heat dissipation and vibration control when compared to existing honeycomb material. Walsh et al. (2017) studied carbon fibre sandwich made from expanded cork (lightweight material). The specimen was exposed to acoustic, impact damage resistance, damping, and bending stiffness tests. As a validation, the developed sandwich was compared with carbon fibre sandwich developed using aerospace standard commercial synthetic foam (Rohacell 110 IG). The results conclude that the developed natural sandwich composite showed superior performance in damping, acoustic and impact damage resistance. However, bending stiffness was relatively lesser than the counterpart. Thus the developed composite can be a potential solution for aerospace structural applications. Chok et al. (2017) explored the possibilities of using NFPC in aircraft radome structures. In this study, three composites are compared namely: kenaf fibre composites, glass fibre composites and hybrid glass reinforced kenaf composites. These composites are subjected to various tests. Finally, the results indicate that the kenaf composites exhibited the least natural frequency. Thus, usage of hybrid composites reduces the impact of destruction over the structure.

3.3 POLYMER NANOCOMPOSITES

Reinforcement of at least one dimension of 1–100 nm filler particles in composite material is known as nanocomposite (Mahenderkar et al. 2017). Fillers are an important part of nano polymer composites. Nearly 50 M tons of fillers are produced every year. These fillers enhance the properties of the polymer composites such as optical, thermal, mechanical, electrical and non-flammability properties, and controlled rheological features (Jabeen et al. 2015). Generally, nanofillers can be classified as

- 1D (nanotubes and nano fibres),
- 2D (shell, laminates) and
- 3D (beads and silica)

The different nanofiller (graphite, CNT, clay and graphite oxide) used in aerospace applications are discussed. Graphite and diamond are carbon allotropes. Generally, carbon occurs in three systems – diamond, fullerenes and graphite. Each of these forms has a different type of carbon bonding (tetrahedral and trigonal hybridization). Moreover, graphene is the carbon layer of graphite. The carbon nanotube (CNT) is a cylindrical pipe with a diameter of 0.4–2.5 nm (Bystrzejewski et al. 2010). The CNT is generally classified into single-walled carbon nanotube (SWCNT) and multi-walled carbon nanotube (MWCNT). Two or more concentric arrangements of SWCNT form MWCNT (Köhler et al. 2008). The addition of CNT has improved the properties of the composites over metal-reinforced composites. This is due to the better compatibility of the CNT with the composite.

Clay is mostly recognised as kaolin (aluminium silicate) and is available in large quantities in nature. In case of polymer nanocomposite, clay minerals such as bentonite, kaolin, mica and talc are widely utilised (Kausar 2017). Composites reinforced with clay showed improvement in electrical insulation, hardness, opacity, brightness and thermal stability (Zhang et al. 2015)

Graphene is more stable and has several advantages which can be suitable for aerospace application. Graphene oxide (GO) is also recognised as functionalised graphene, chemically modified graphene and reduced grapheme (Rafiee 2011).

Some of the literatures related to polymer nanocomposite are tabulated in Table 3.1

3.4 ANTI-STATIC POLYMERS

The accumulation of static electricity was first witnessed by Thales of Miletus (624 BC–547 BC). He explained that amber appeals to smaller particles when rubbed over animal fur. Further, William Gilbert (1544–1603) explored that there were a few other materials that can develop static electricity apart from amber. The electrostatic charge developed is not a stable charge, neither repulsive nor attractive, but it will dissipate over time depending on few parameters (Yadav et al. 2020). Generally,

TABLE 3.1

Some Literatures Related to Polymer Nanocomposite Used in Aerospace Application

S. No.	Authors	Nanocomposite	Description
1	Zhou et al. (2013)	Polyamide / graphite	Investigations on polyamide reinforced with graphite were conducted. Various compositions of graphites are mixed with polyamide and samples are prepared. An improved thermal conductivity occurred for the samples that had graphite content less than 30 wt.%. Subsequently, the electrical, mechanical and thermal properties of the composite have been improved by the reinforcement of graphite.
2	Lee et al. 2016	Polymer / CNT	The MWCNT was blended with multiple hydrogen bonding moiety (MHB) and the composite was developed and various tests were conducted. The results indicate that the tensile strength increased from 156 MPa to 278 MPa, strain increased from 3.7% to 7%, modulus from 11.3 GPa to 13.9 GPa and electrical conductivity from 10^{-8} S/cm to 10^{0} S/cm.
3	Xu et al. 2008	Polymer / CNT	The polypropylene was bent with covalently functionalised CNT. The diameter of the CNT was maintained within 20–90 nm. The CNT improved the flame retardancy and mechanical properties of polypropylene.
4	Guo et al. 2018	Polymer / nanoclay	Investigation on polymer/clay nanocomposite was conducted. Results conclude that the composite increases the barrier features, mechanical strength high-temperature resistance, and decreases flammability.
5	Kausar, Rafique, and Muhammad (2017).	Polymer / graphene oxide	The epoxy was reinforced with GO. The composite was subjected to various tests. The results reveal that the improvements are observed in Young's modulus, tensile strength and load transfer due to GO addition. Additionally, the tensile strength improved from 7 MPa to 13 MPa (1.5 vol%). On further increasing the GO ratio, a decline in tensile strength (7.5 MPa) was noted.
6	Song et al. 2011	Polymer / graphene oxide	In this work, polypropylene was blended with exfoliated graphene. The addition of 0.42 vol% of graphene leads to an increase in Young's modulus by 74% and yield strength by 75% in the composite.

based on the electrical characterisation, the materials (Wypych and Pionteck 2016) are categorised as

- Insulator ($\sigma = 10^{-21}$ to 10^{-11} S·cm^{-1}),
- Semiconductor ($\sigma = 10^{-10}$ to 10^{-3} S·cm^{-1}),
- Conductors ($\sigma = 10^3$ S·cm^{-1}) and
- Superconductors ($\sigma = 10^{21}$ S·cm^{-1})

Moreover, the relative permittivity, volume and surface resistivity are terms to describe the insulating properties of a material as per the ASTM standards. On considering this, most of the polymers come under the insulator category with $\sigma = 10^{-13}$ to 10^{-22} S·cm^{-1}. Since electrons in the polymers are localised in a vitrified bond (no free electrons), the polymers behave as an insulator (Yuan et al. 2018). Presently, the usage of polymer composites is increasing exponentially in aerospace applications. Therefore, conductivity is an important parameter to be taken into account while considering static electricity.

Aircraft parts are subjected to various interactions like with air, relative motion between components, flaps and rudder movements, and so on. Thus, there are changes for static charge accumulation which may lead to damage of aircraft electric components (Gigliotti 2011).

Consequently, Figure 3.3 illustrates the static charge over the aircraft (Giacometti and Oliveira 1992).

The conventional materials used in aircraft (aluminium skin) provide conductive paths during a lightning strike. However, in the conventional materials, the accumulation of static charge is experienced, which may lead to severe damage to aircraft (Raimondo 2014). The invention of nanotechnology for polymer composites has surpassed the ability of polymer composites in terms of better handling of static charge, improving conductivity and enhancing other properties of composites (Pramanik et al. 2013). Even though polymers are insulators, addition of nanofilllers (carbon fibre, carbon black (CB), metal particles, and graphene) have showed significant increase in conductivity (Jeong et al. 2009).

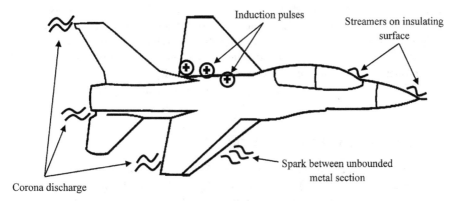

FIGURE 3.3   Static electricity over the aircraft.

TABLE 3.2
Electrical Properties of Some of the Polymer Nanocomposite

Filler	Ploymers	CNT (wt%)	Percolation threshold (wt%/vol%)	Composite conductivity (S.m^{-1})	Authors
MWCNT	PA	~12	–	0.01	Yuan et al. 2009
MWCNT	PU	< 27	0.009	2×10^3	Koerner et al. 2005
Oxidised MWCNT	Epoxy	< 1	0.012	1×10^{-2}	Spitalský et al. 2009
Pulmatic acid modified CNT	Epoxy	< 0.8	0.05–0.1	6.9×10^{-3}	Barrau et al. 2003
SWCNT	Epoxy	< 15	0.062	10	Cui et al. 2003
MWCNT	Epoxy	< 0.18	–	4×10^{-1}	Sandler et al. 1999

CB is developed by the partial ignition of hydrocarbon (aromatic) at higher temperature (Huang 2002). In aerospace applications, CB is the most widely used nanofiller in epoxy resin as an antistatic agent. CB is a conductive nanofiller and ultra violet light stabiliser. The usage of CB forms a three dimensional network which insulates and also supports conductivity, which is known as percolation threshold (Macutkevic et al. 2013). The effect of CB reinforcing in epoxy was explored by Aal et al. (2008). The results revealed that, by maintain the weight fraction of CB $\leq$ 15wt%, the surface resistivity showed in the series of $1.62 \times 10^6 - 1.1 \times 10^5$ Ω·cm.

The graphene incorporation in the composite has improved the conductivity by 10^6 S·m^{-1} (Wang et al. 2017), which can be a suitable solution for aerospace applications. Consequently, Liang et al. (2009) illustrated the usage of graphene in the epoxy-based composite and reported that the percolation threshold was relatively low. The measured conductivity was 0.05 S·cm^{-1} and percolation threshold was 0.5 vol% for 8.8 vol% of graphene content.

Gagné and Therriault (2014) gave a detailed explanation over a different approach to develop the composite that performs better against static charge and lightning. In recent years, onion-like carbon (OLC) is utilised for increasing the electricity of the composites. Consequently, Palaimiene et al. (2018) developed an epoxy-based composite with incorporation of OLC. The results demonstrate that the 250 nm OLC was able to produce 0.7 vol% (lowest percolation threshold) for conductivity. Additionally, Zeiger et al. (2016) consolidated a detailed review on nano-diamonds–derived OLC. The properties of the different OLC were investigated.

Table 3.2 shows the electrical properties of some of the polymer nanocomposite (Spitalsky et al. 2010)

3.5   ELECTRO ACTIVE POLYMERS

The polymers that respond mechanically to electrical stimulation are known as electro active polymers (EAPs) (Bashir & Rajendran 2018). EAP can function as both sensor and also as actuator. This phenomenon is due to the electromechanical property of the EAP (Jo et al. 2013). A significant change in shape was observed when

EAP was exposed to electricity. Further, the maximum strain reached was 300% (Bar-Cohen 2017). The EAP is widely utilised as an actuator due to high actuation strain, numerous actuation modes and biomimetic ability (Tondu 2007). Moreover, EAP are anticipated to be compact, light and to consume less energy (Asaka and Nakamura, 2019). Both of these requirements are necessary for NASA to build morphing unmanned aerial vehicles (UAVs), powerful wings, robotic arms and other advanced technologies (Bar-Cohen 2000).

The enactment of multiple ionic EAP was analysed by Punning et al. (2014). The outcomes indicate that the i-EAP was able to withstand the X-ray radiations and ionizing gamma in space. Consequently, Jo et al. (2013) investigated i-EAP constructed using flemion and nafion. The developed i-EAP can be potential solution to the actuator and sensor materials. Further, a flexible tactile sensor was developed using the flemion ion composites by Wang et al. 2009. The i-EAP actuators can be promising solution to the MEMS and smart materials since it has low operating voltage and larger strain (Li et al. 2011)

Ren et al. (2016) developed a hybrid actuator made of shape memory alloy and EAP, and the thermal response of the hybrid was investigated. The results showed that the ratio of EAP and shape memory alloy play a vital role in the actuator displacement. Further, promising results were obtained which indicate that the developed hybrid actuator can be used in space and morphing applications. Li et al. (2017) developed origami-inspired artificial muscles. Theses muscles are fluid-driven and can generate stress up to 3.6 MPa, contract 90% from their original length and produce 2kW/kg of peak power density. They can be employed in space applications due to low cost fabrication and robustness. Furthermore, visco-elastic EAP (VEAP) can provide better shock and vibration damping, more flexible actuator and realistic aesthetics (Nguyen et al. 2014). There is a huge scope to explore the possibilities of EAP in the aerospace applications.

3.6 SUMMARY

From the present chapter, the following major findings with respect to the polymer composites for aerospace application are been derived

- The low-cost and high-strength polymer composites can be a suitable solution for aerospace parts (interior and exterior) than conventional alloys.
- The increase in demand for the fibre materials can be compensated by using natural fibres. The developed NFPCs can be a potential solution for interior aircraft parts since they showed significant performance when compared with the existing composites.
- Various properties of polymer composites such as optical, mechanical and thermal properties, as well as corrosion resistance and conductivity are increased by incorporating nanofillers. Thus nano polymer composites can be used in aerospace application due to their phenomenal properties.
- Electroactive polymers are low cost and consume less energy than the conventional actuators. Thus electroactive polymers can replace the existing actuators in aerospace applications.

Therefore from these major findings, polymer composites can be a promising solution for various parts of the aerospace application.

REFERENCES

Aal NA, El-Tantawy F, Al-Hajry A, Bououdina M. New antistatic charge and electromagnetic shielding effectiveness from conductive epoxy resin/plasticized carbon black composites. Polymer Composites. 2008 Feb;29(2):125–32.

Asaka K, Nakamura K. Current status of applications and markets of soft actuators. In Soft actuators 2019 (pp. 19–35). Springer, Singapore.

Asim M, Saba N, Jawaid M, Nasir M. Potential of natural fibre/biomass filler-reinforced polymer composites in aerospace applications. In Sustainable composites for aerospace applications 2018 Jan 1 (pp. 253–268). Woodhead Publishing.

Bar-Cohen Y. Electroactive polymers as artificial muscles: capabilities, potentials and challenges. In Robotics 2000 2000 (pp. 188–196).

Bar-Cohen Y. Humanoids and the potential role of electroactive materials/mechanisms in advancing their capability. In Advances in science and technology 2017 (Vol. 97, pp. 81–89). Trans Tech Publications Ltd.

Barrau S, Demont P, Perez E, Peigney A, Laurent C, Lacabanne C. Effect of palmitic acid on the electrical conductivity of carbon nanotubes– epoxy resin composites. Macromolecules 2003 Dec 30;36(26):9678–80.

Bashir M, Rajendran P. A review on electroactive polymers development for aerospace applications. Journal of Intelligent Material Systems and Structures. 2018 Nov;29(19):3681–95.

Boegler O, Kling U, Empl D, Isikveren AT. Potential of sustainable materials in wing structural design. Deutsche Gesellschaft für Luft-und Raumfahrt-Lilienthal-Oberth eV; 2015.

Bystrzejewski M, Rummeli MH, Gemming T, Lange H, Huczko A. Catalyst-free synthesis of onion-like carbon nanoparticles. New Carbon Materials. 2010 Feb 1;25(1):1–8.

Chok EY, Majid DL, Harmin MY. Effect of low velocity impact damage on the natural frequency of composite plates. *InIOP Conference Series: Materials Science and Engineering* 2017 Dec 1 (Vol. 270, No. 1, p. 012025). IOP Publishing.

Cui S, Canet R, Derre A, Couzi M, Delhaes P. Characterization of multiwall carbon nanotubes and influence of surfactant in the nanocomposite processing. Carbon. 2003 Jan 1;41(4):797–809.

Eloy FS, Costa RR, De Medeiros R, Ribeiro ML, Tita V. Comparison between mechanical properties of bio and synthetic composites for use in aircraft interior structures. In Meeting on aeronautical composite materials and structures, São Carlos, Brazil 2015.

Gagné M, Therriault D. Lightning strike protection of composites. Progress in Aerospace Sciences. 2014 Jan 1;64:1–6.

Ghori SW, Siakeng R, Rasheed M, Saba N, Jawaid M. The role of advanced polymer materials in aerospace. In Sustainable composites for aerospace applications 2018 Jan 1 (pp. 19–34). Woodhead Publishing.

Giacometti JA, Oliveira ON. Corona charging of polymers. IEEE Transactions on Electrical Insulation. 1992 Oct;27(5):924–43.

Gigliotti K. Static electricity and aircraft. Wiley encyclopedia of composites. 2011 Sep 16:1–8.

Guo F, Aryana S, Han Y, Jiao Y. A review of the synthesis and applications of polymer–nanoclay composites. Applied Sciences. 2018 Sep;8(9):1696.

Gupta MC. Polymer composite. New Age International; 2007.

Huang JC. Carbon black filled conducting polymers and polymer blends. Advances in Polymer Technology: Journal of the Polymer Processing Institute. 2002 Dec;21(4):299–313.

Huda MS, Drzal LT, Mohanty AK, Misra M. Chopped glass and recycled newspaper as reinforcement fibres in injection molded poly (lactic acid)(PLA) composites: a comparative study. Composites Science and Technology. 2006 Sep 1;66(11–12):1813–24.

Irving PE, Soutis C, editors. Polymer composites in the aerospace industry. Woodhead Publishing; 2019 Nov 26.

Jabeen S, Kausar A, Muhammad B, Gul S, Farooq M. A review on polymeric nanocomposites of nanodiamond, carbon nanotube, and nanobifiller: Structure, preparation and properties. Polymer-Plastics Technology and Engineering. 2015 Sep 17;54(13):1379–409.

Jeong MY, Byung-Yoon AH, Sang-Koul LE, Won-Ki LE, Nam-Ju JO. Antistatic coating material consisting of poly (butylacrylate-co-styrene) core-nickel shell particle. Transactions of Nonferrous Metals Society of China. 2009 Sep 1;19:s119–23.

Jo C, Pugal D, Oh IK, Kim KJ, Asaka K. Recent advances in ionic polymer–metal composite actuators and their modeling and applications. Progress in Polymer Science. 2013 Jul 1;38(7):1037–66.

Joshi M, Chatterjee U. Polymer nanocomposite: an advanced material for aerospace applications. In Advanced composite materials for aerospace engineering 2016 Jan 1 (pp. 241–264). Woodhead Publishing.

Kausar A, Rafique I, Muhammad B. Aerospace application of polymer nanocomposite with carbon nanotube, graphite, graphene oxide, and nanoclay. Polymer-Plastics Technology and Engineering. 2017 Sep 2;56(13):1438–56.

Kausar A. Physical properties of hybrid polymer/clay composites. In Hybrid polymer composite materials 2017 Jan 1 (pp. 115–132). Woodhead Publishing.

Koerner H, Liu W, Alexander M, Mirau P, Dowty H, Vaia RA. Deformation–morphology correlations in electrically conductive carbon nanotube—thermoplastic polyurethane nanocomposites. Polymer. 2005 May 26;46(12):4405–20.

Köhler AR, Som C, Helland A, Gottschalk F. Studying the potential release of carbon nanotubes throughout the application life cycle. Journal of Cleaner Production. 2008 May 1;16(8–9):927–37.

Ku H, Wang H, Pattarachaiyakoop N, Trada M. A review on the tensile properties of natural fibre reinforced polymer composites. Compos Part B 2011;42:856–873.

Lee TW, Han M, Lee SE, Jeong YG. Electrically conductive and strong cellulose-based composite fibres reinforced with multiwalled carbon nanotube containing multiple hydrogen bonding moiety. Composites Science and Technology. 2016 Feb 8;123:57–64.

Li J, Ma W, Song L, Niu Z, Cai L, Zeng Q, Zhang X, Dong H, Zhao D, Zhou W, Xie S. Superfast-response and ultrahigh-power-density electromechanical actuators based on hierarchal carbon nanotube electrodes and chitosan. Nano letters. 2011 Nov 9;11(11):4636–4641.

Li S, Vogt DM, Rus D, Wood RJ. Fluid-driven origami-inspired artificial muscles. Proceedings of the National academy of Sciences. 2017 Dec 12;114(50):13132–13137.

Liang J, Wang Y, Huang Y, Ma Y, Liu Z, Cai J, Zhang C, Gao H, Chen Y. Electromagnetic interference shielding of graphene/epoxy composites. Carbon. 2009 Mar 1;47(3):922–5.

Macutkevic J, Kuzhir P, Paddubskaya A, Maksimenko S, Banys J, Celzard A, Fierro V, Bistarelli S, Cataldo A, Micciulla F, Bellucci S. Electrical transport in carbon black-epoxy resin composites at different temperatures. Journal of Applied Physics. 2013 Jul 21;114(3):033707.

Mahenderkar NK, Prabhu TR, Kumar A. Nanocomposites potential for aero applications. In Aerospace materials and material technologies 2017 (pp. 391–411). Springer, Singapore.

Manan NH, Majid DL, Romli FI. Mould design and manufacturing considerations of honeycomb biocomposites with transverse fibre direction for aerospace application. In *IOP Conference Series: Materials Science and Engineering* 2016 Oct 1 (Vol. 152, No. 1, p. 012013). IOP Publishing.

Mansor MR, Nurfaizey AH, Tamaldin N, Nordin MN. Natural fibre polymer composites: utilization in aerospace engineering. In Biomass, biopolymer-based materials, and bioenergy 2019 Jan 1 (pp. 203–224). Woodhead Publishing.

Nguyen CH, Alici G, Mutlu R. Modeling a soft robotic mechanism articulated with dielectric elastomer actuators. In *2014 IEEE/ASME International Conference on Advanced Intelligent Mechatronics* 2014 Jul 8 (pp. 599–604). IEEE.

Palaimiene E, Macutkevic J, Banys J, Selskis A, Fierro V, Celzard A, Schaefer S, Shendcrova O. Ultra-low percolation threshold in epoxy resin–onion-like carbon composites. Applied Physics Letters. 2018 Jul 16;113(3):033105.

Pramanik S, Hazarika J, Kumar A, Karak N. Castor oil based hyperbranched poly (ester amide)/polyaniline nanofibre nanocomposites as antistatic materials. Industrial & Engineering Chemistry Research. 2013 Apr 24;52(16):5700–7.

Punning A, Kim KJ, Palmre V, Vidal F, Plesse C, Festin N, Maziz A, Asaka K, Sugino T, Alici G, Spinks G. Ionic electroactive polymer artificial muscles in space applications. Scientific Reports. 2014 Nov 5;4(1):1–6.

Rafiee MA. Graphene-based composite materials. New York, USA: Rensselaer Polytechnic Institute; 2011 May.

Raimondo, M. Improving the aircraft safety by advanced structures and protecting nanofillers. 2014.

Red C. Composites in aircraft interiors, 2012–2022. Composites World. 2012 Sep.

Ren K, Bortolin RS, Zhang QM. An investigation of a thermally steerable electroactive polymer/shape memory polymer hybrid actuator. Applied Physics Letters. 2016 Feb 8; 108(6):062901.

Saba N, Paridah MT, Jawaid M, Abdan K, Ibrahim NA. Manufacturing and processing of kenaf fibre-reinforced epoxy composites via different methods. In Manufacturing of natural fibre reinforced polymer composites 2015 (pp. 101–124). Springer, Cham.

Sandler J, Shaffer MS, Prasse T, Bauhofer W, Schulte K, Windle AH. Development of a dispersion process for carbon nanotubes in an epoxy matrix and the resulting electrical properties. Polymer. 1999 Oct 1;40(21):5967–71.

Song P, Cao Z, Cai Y, Zhao L, Fang Z, Fu S. Fabrication of exfoliated graphene-based polypropylene nanocomposites with enhanced mechanical and thermal properties. Polymer. 2011 Aug 18;52(18):4001–10.

Špitalský Z, Krontiras CA, Georga SN, Galiotis C. Effect of oxidation treatment of multiwalled carbon nanotubes on the mechanical and electrical properties of their epoxy composites. Composites Part A: Applied Science and Manufacturing. 2009 Jul 1;40(6–7):778–83.

Spitalsky Z, Tasis D, Papagelis K, Galiotis C. Carbon nanotube–polymer composites: chemistry, processing, mechanical and electrical properties. Progress in Polymer Science. 2010 Mar 1;35(3):357–401.

Tondu B. Artificial muscles for humanoid robots. ed; 2007 Jun 1.

Walsh J, Kim HI, Suhr J. Low velocity impact resistance and energy absorption of environmentally friendly expanded cork core-carbon fibre sandwich composites. Composites Part A: Applied Science and Manufacturing. 2017 Oct 1;101:290–6.

Wang J, Ma F, Sun M. Graphene, hexagonal boron nitride, and their heterostructures: properties and applications. RSC Advances. 2017;7(27):16801–22.

Wang J, Sato H, Xu C, Taya M. Bioinspired design of tactile sensors based on Flemion. Journal of Applied Physics. 2009 Apr 15;105(8):083515.

Wypych G, Pionteck J. Handbook of antistatics. Elsevier; 2016 Oct 3.

Xu L, Guo Z, Zhang Y, Fang Z. Flame-retardant-wrapped carbon nanotubes for simultaneously improving the flame retardancy and mechanical properties of polypropylene. Journal of Materials Chemistry. 2008;18(42):5083–91.

Yadav R, Tirumali M, Wang X, Naebe M, Kandasubramanian B. Polymer composite for antistatic application in aerospace. Defence Technology. 2020 Feb 1;16(1):107–18.

Yuan S, Zheng Y, Chua CK, Yan Q, Zhou K. Electrical and thermal conductivities of MWCNT/polymer composites fabricated by selective laser sintering. Composites Part A: Applied Science and Manufacturing.s 2018 Feb 1;105:203–13.

Yuan WZ, Lam JW, Shen XY, Sun JZ, Mahtab F, Zheng Q, Tang BZ. Functional polyacetylenes carrying mesogenic and polynuclear aromatic pendants: polymer synthesis, hybridization with carbon nanotubes, liquid crystallinity, light emission, and electrical conductivity. Macromolecules. 2009 Apr 14;42(7):2523–31.

Zeiger M, Jäckel N, Mochalin VN, Presser V. carbon onions for electrochemical energy storage. Journal of Materials Chemistry A. 2016;4(9):3172–96.

Zhang Y, Stringer J, Grainger R, Smith PJ, Hodzic A. Fabrication of patterned thermoplastic microphases between composite plies by inkjet printing. Journal of Composite Materials. 2015 Jun;49(15):1907–13.

Zhou S, Chen Y, Zou H, Liang M. Thermally conductive composites obtained by flake graphite filling immiscible polyamide 6/polycarbonate blends. Thermochimica Acta. 2013 Aug 20;566:84–91.

4 The Mechanical Properties of Kenaf/ Jute Hybrid Composites in Different Angles of Orientation

K. Tabrej

Universiti Putra Malaysia, Serdang, Malaysia

M. T. H. Sultan

Universiti Putra Malaysia, Serdang, Malaysia

Aerospace Malaysia Innovation Centre (944751-A), Cyberjaya, Malaysia

A.U.M. Shah

Universiti Putra Malaysia, Serdang, Malaysia

S.N.A. Safri

Universiti Putra Malaysia, Serdang, Malaysia

M. Jawaid

Universiti Putra Malaysia, Serdang, Malaysia

CONTENTS

DOI: 10.1201/9781003200994-4

4.1   INTRODUCTION

Hybrid natural fibres with other natural or synthetic fibres are one of the solutions to enhance the properties of composites. The choice of fibres in hybrid composites depends on the needs of the end product to avoid an overdesign, which can result in extra weight and cost. Natural-natural hybrid composites generally offer lower density, and thus are lighter in weight compared to natural-synthetic hybrid composites [1, 2]. In achieving more environmentally friendly products, natural-natural hybrid composites are also a better choice, with comparable strength properties [3, 4]. There are rising number of products made of natural fibres in the market nowadays, which confirms its reliability in different applications [5].

The mechanical properties of composites depends on many factors, such as length, types and sizes of fibres, chemical composition of fibres, percentage of fibre loading in composites, surface modifications on fibres and orientation of fibres in composites [6]. Compared to short fibres in powder or single fibre types, woven natural fibres offer better strength and mechanical properties as reinforcements [4, 7]. However there are some limitations, in which the woven fibre sometime cannot fit the mould of the product perfectly, thus short fibre is the best alternative [8, 9].

In the current study, the woven kenaf and jute fibres were used as reinforcements in epoxy matrix. The composites were prepared in different angles of orientation to observe their mechanical performances through tensile, flexural, compression and Charpy impact tests. Both woven fibres were treated with sodium hydroxide to enhance their compatibility in the polymer matrix. The fractured surfaces of composites were later examined by scanning electron microscope.

4.2   METHODOLOGY

Kenaf and jute fibres used in the current study were both in woven forms, with thickness approximately 0.7–0.8mm, and supplied by Indersen Shamlal Pvt. Ltd. India. The density of kenaf and jute fibre are 1.4 gm/cm^3 and 1.13gm/cm^3 respectively. The matrix used was epoxy resin from the brand Smooth-On, purchased from Mecha Solve Engineering Sdn. Bhd, Selangor, Malaysia. The properties of kenaf and jute fibres are given in Table 4.1.

4.2.1   FABRICATION K/J/K HYBRID COMPOSITES

The hybrid composites were fabricated in constant layering sequence of K/J/K through hand lay-up method. The composites were prepared with 30:70 fibre to

TABLE 4.1

Properties of Kenaf and Jute Fibres [1]

Properties	Kenaf fibres	Jute fibres
Density (g/cc)	1.4	1.13
Tensile strength (MPa)	295–1190	393–773
Modulus (GPa)	20–60	10–30
Elongation (%)	1.75–1.9	1.15–1.5
Cellulose content (%)	53.57	61–70
Hemi cellulose (%)	15–19	13.6–16
Lignin content (%)	5–11	12–13

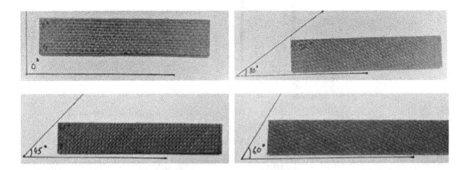

FIGURE 4.1 Hybrid composites in different angles of orientation.

matrix ratio and cured at room temperature. The cured composites were then cut into different angles of orientation, 0°, 30°, 45° and 60°, as shown in Figure 4.1.

4.2.2 TENSILE TESTING

The tensile test was performed according to ASTM D 3039 standard. The dimensions of sample are 120 mm × 20 mm × 3 mm for length, width and thickness, respectively. The samples were gripped by the tensile jig and then pulled until the specimen failed in room temperature. The crosshead speed rate was set to 2 mm/min.

4.2.3 FLEXURAL TESTING

The flexural test was carried out in a three-point bending mode, following the ASTM D7264 standard. The dimensions of sample are 120 mm × 15 mm × 3.2 mm for length, width and thickness. The crosshead speed is 1.5 mm/min. The span length was set to 100 mm.

4.2.4 SCANNING ELECTRON MICROSCOPY (SEM)

The morphological images of the hybrid composites specimen from the tensile and flexural testing was analysed using the Coxem EM-30AX Scanning Electron

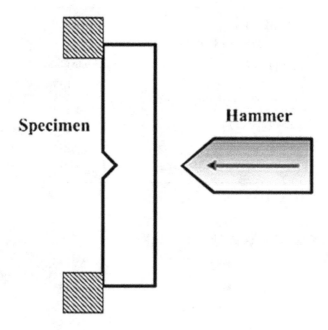

FIGURE 4.2   Set up for Charpy impact test.

Microscopy (SEM), with emission current of 93000 nA and the applied voltage of 10.0 kV. The working distance was fixed to 8300 um.

4.2.5 Charpy Impact Testing

Charpy impact test was carried out to evaluate the impact toughness of the hybrid composites. The dimensions of composites were 65 mm × 15 mm × 3.12 mm for length, width and thickness, with a V-notched as shown in Figure 4.2. The test was conducted as per ASTM D256 (Instron, Norwood, United States) standard.

4.3 RESULTS AND DISCUSSION

4.3.1 Tensile Properties

The strength of K/J/K hybrid composites depends on the woven mat fibres' orientation and the bonding between the fibre and matrix. As shown in Figure 4.3, the tensile strength and tensile modulus of K/J/K (0°) hybrid composites is the highest among other orientations, K/J/K (30°), (45°) and (60°). This could be due to an even spreading of stress transfer with the application of tensile load in both the longitudinal and transverse directions [10, 11].

The tensile strength of K/J/K hybrid composites is attributed to differences in the load-sharing properties of kenaf and jute fibres along the longitudinal and transverse directions. Tensile strength is higher when woven mat fibres orientation (0°) is expected based on previous reported studies [12, 13].

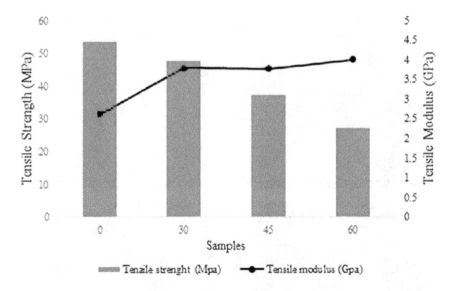

FIGURE 4.3 Tensile strength and modulus of K/J/K hybrid composites in different angel of orientations 0°, 30°, 45° and 60°.

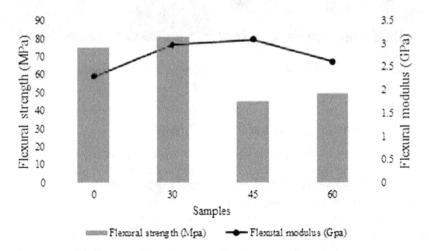

FIGURE 4.4 Flexural strength and modulus of K/J/K hybrid composites in different angle of orientations 0°, 30°, 45° and 60°.

4.3.2 FLEXURAL PROPERTIES

The flexural strength and modulus of K/J/K hybrid composites are shown in Figure 4.4 Slightly different results were found compared to the tensile strength, which the K/J/K (30°) showed as the highest flexural strength. However, there was not much difference in value compared to K/J/K (0°) as the second-highest flexural strength. The low flexural strength of K/J/K (45°) and K/J/K (60°) suggests poor load bearing capability, as the length of fibre is shorter from end to end to meet the angle limitations [14].

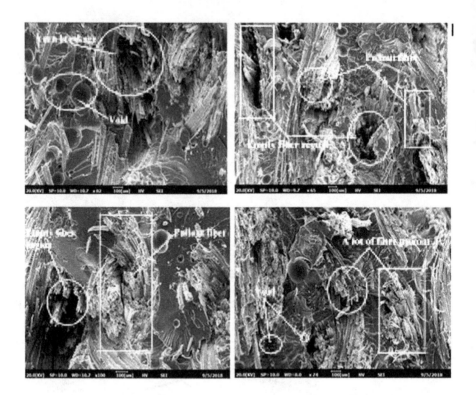

FIGURE 4.5   SEM images of K/J/K hybrid composites from the tensile fractured samples.

4.3.3   SCANNING ELECTRON MICROSCOPY (SEM)

The SEM images of the fracture samples from the tensile testing are presented in Figure 4.5. Several types of fracture characteristics were observed, such as fibre fracture, fibre pull out, matrix cracking, fibre de-bonding and voids in matrix.

It can be seen that void existence is one of the factor for K/J/K composites failure. Poor fabrication method attributed to more void existence in composites. Besides that, the empty fibre region suggested that the fibres were not fully penetrated with the epoxy during the hand lay-up process. This may result in low stress transfer, and thus low mechanical strength [15].

4.3.4   IMPACT PROPERTIES

Figure 4.6 shows the impact properties of K/J/K hybrid composites with different orientation, 0°, 30°, 45°, and 60°. The impact energy of K/J/K (0°) is 6170.48 J/m^2, K/J/K (30°) is 6157.64 J/m^2, K/J/K (45°) is 5573.68 J/m^2 and K/J/K (60°) is 53.92.54 J/m^2.

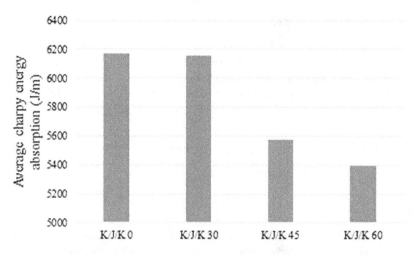

FIGURE 4.6 Energy absorbed by K/J/K hybrid composites through the Charpy impact testing.

A low interfacial bonding between the kenaf/jute fiber and the epoxy matrix allows for easier decohesion during the impact, which causes fracture by delamination. This mechanism of fracture is more effective in composites with greater strength, K/J/K (0°) and K/J/K (30°). Higher energy was required to break the strength and initiate cracks for composites with good fibre-matrix bonding.

4.4 CONCLUSION

The current research deals with the development of kenaf/jute hybrid composites, fabricated through conventional hand lay-up process. Constant sequence of K/J/K was selected based on previous studies and the angle of orientation was varied at 0°, 30°, 45°, and 60°. The developed kenaf/jute hybrid composites were evaluated for their mechanical properties as per ASTM standards and the morphological observation was carried out by Scanning Electron Microscopy. From this research work, the following conclusions are made:

1. The highest tensile strength was presented in K/J/K (0°) at 53 MPa, while the lowest was shown in K/J/K (60°) at 27 MPa.
2. The highest flexural strength was presented in K/J/K (30°) at 80 MPa, while the lowest was shown in K/J/K (45°) at 46 MPa.
3. There were only slight differences between K/J/K (0°) and K/J/K (30°) for the amount of energy absorbed through the impact testing, which both marked a high value approximately at 6200 (J/m).
4. The images from the SEM analysis suggested that void existence and the poor wettability of fibres are the major contributions to composites failure.

ACKNOWLEDGEMENTS

This work was supported by Universiti Putra Malaysia under Geran Putra Berimpak, GPB 9668200. The authors would like to express their gratitude and sincere appreciation to the Department of Aerospace Engineering and Laboratory of Biocomposite Technology, Institute of Tropical Forestry and Forest Products (INTROP), Universiti Putra Malaysia, UPM (HICOE) for the close collaboration in this research.

DECLARATION OF INTEREST STATEMENT

The authors declared no conflict of interest.

REFERENCES

1. T. Khan, M. T. B. Hameed Sultan, and A. H. Ariffin, "The challenges of natural fiber in manufacturing, material selection, and technology application: A review," *Journal of Reinforced Plastics and Composites*, vol. 37, no. 11, pp. 770–779, 2018.
2. R. Yahaya, S. Sapuan, M. Jawaid, Z. Leman, and E. Zainudin, "Effect of layering sequence and chemical treatment on the mechanical properties of woven kenaf–aramid hybrid laminated composites," *Materials & Design*, vol. 67, pp. 173–179, 2015.
3. M. Jawaid, O. Y. Alothman, M. Paridah, and H. S. A. Khalil, "Effect of oil palm and jute fiber treatment on mechanical performance of epoxy hybrid composites," *International Journal of Polymer Analysis and Characterization*, vol. 19, no. 1, pp. 62–69, 2014.
4. M. Jawaid, H. A. Khalil, A. Hassan, R. Dungani, and A. Hadiyane, "Effect of jute fibre loading on tensile and dynamic mechanical properties of oil palm epoxy composites," *Composites Part B: Engineering*, vol. 45, no. 1, pp. 619–624, 2013.
5. R. Nadlene, S. Sapuan, M. Jawaid, M. Ishak, and L. Yusriah, "A review on roselle fiber and its composites," *Journal of Natural Fibers*, vol. 13, no. 1, pp. 10–41, 2016.
6. M. Sanjay, G. Arpitha, and B. Yogesha, "Study on mechanical properties of natural-glass fibre reinforced polymer hybrid composites: A review," *Materials Today: Proceedings*, vol. 2, no. 4–5, pp. 2959–2967, 2015.
7. F. M. Al-Oqla, S. Sapuan, T. Anwer, M. Jawaid, and M. Hoque, "Natural fiber reinforced conductive polymer composites as functional materials: A review," *Synthetic Metals*, vol. 206, pp. 42–54, 2015.
8. M. Z. Rahman, "Mechanical performance of natural/natural fiber reinforced hybrid composite materials using finite element method based micromechanics and experiments," Thesis, Master of Science, Utah State University, 2017.
9. M. J. Sharba, Z. Leman, M. T. Sultan, M. R. Ishak, and M. A. A. Hanim, "Effects of kenaf fiber orientation on mechanical properties and fatigue life of glass/kenaf hybrid composites," *BioResources*, vol. 11, no. 1, pp. 1448–1465, 2015.
10. P. Amuthakkannan, V. Manikandan, J. T. W. Jappes, and M. Uthayakumar, "Hybridization effect on mechanical properties of short basalt/jute fiber-reinforced polyester composites," *Science and Engineering of Composite Materials*, vol. 20, no. 4, pp. 343–350, 2013.
11. K. Senthilkumar et al., "Static and dynamic properties of sisal fiber polyester composites–effect of interlaminar fiber orientation," *BioResources*, vol. 12, no. 4, pp. 7819–7833, 2017.
12. G. Uzun, N. Hersek, and T. Tincer, "Effect of five woven fiber reinforcements on the impact and transverse strength of a denture base resin," *The Journal of Prosthetic Dentistry*, vol. 81, no. 5, pp. 616–620, 1999.

13. S. D. Salman, Z. Leman, M. T. Sultan, M. R. Ishak, and F. Cardona, "The effects of orientation on the mechanical and morphological properties of woven kenaf-reinforced poly vinyl butyral film," *BioResources*, vol. 11, no. 1, pp. 1176–1188, 2015.
14. R. Yahaya, S. Sapuan, M. Jawaid, Z. Leman, and E. Zainudin, "Effect of fibre orientations on the mechanical properties of kenaf–aramid hybrid composites for spall-liner application," *Defence Technology*, vol. 12, no. 1, pp. 52–58, 2016.
15. H. Aisyah et al., "Effects of fabric counts and weave designs on the properties of laminated woven kenaf/carbon fibre reinforced epoxy hybrid composites," *Polymers*, vol. 10, no. 12, p. 1320, 2018.

5 Dynamic Behaviour of Adhesively Bonded Structures in Aerospace Applications
An Overview

Thulasidhas Dhilipkumar and Murugan Rajesh
Vellore Institute of Technology, Vellore, India

Soundhar A
Sri Venkateswara College of Engineering (SVCE), Chennai, India

CONTENTS

5.1 INTRODUCTION

Fibre-reinforced polymer composites are replacing conventional materials such as steel and aluminium in weight-sensitive applications such as aerospace, shipbuilding, robotics, automotive, military and underwater vehicles due to their high strength-to-weight ratio, exceptional stiffness and superior fatigue resistance (Ashcroft, Hughes, and Shaw 2006). However, in most of these applications, the large-scale composite parts are assembled from several subcomponents (Kupski and Teixeira de Freitas 2021). Therefore, joint design plays a major role in the performance of lightweight composite structures because joints are the weakest part of the composite structure.

The composite parts used in aerospace industries can be assembled in two commonly used joining methods, mechanical fastening and adhesive bonding, as illustrated in Figure 5.1. Mechanical fastening has been the popular method for joining

DOI: 10.1201/9781003200994-5

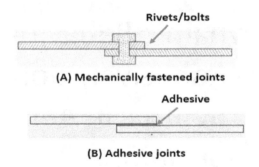

FIGURE 5.1   Common joining technique in aerospace industries.

TABLE 5.1

Advantages and Disadvantages of Adhesively Bonded Joints

S.No	Advantages	Disadvantages
1	• Reduces stress concentration and provides uniform stress distribution.	• Requires careful surface preparation of adherends.
2	• Superior fatigue resistance.	• The selection of appropriate adhesive is important.
3	• Provides excellent sealability and resistance against corrosion.	• Difficult for inspection and repair.
4	• Lightweight structures.	• Joint quality depends on several parameters.
5	• Suitable for the multi-material bonding process.	• Require long curing time.

metallic structures due to the ease of assembling and dismantling. Also, it requires very little surface preparation and it's easy to inspect the quality of joints. However, fastening holes on composite parts causes peeling of laminate plies, severe matrix damage, delamination failure and initiates local stress concentration near the hole area. These damages can induce cracks and lower the fatigue strength of the composite structure (Heidary, Karimi, and Minak 2018). To overcome the disadvantages allied with mechanical fastening, researchers developed an adhesive bonding technique to join composite parts. Adhesive bonding is an appropriate method to join composite parts where the adhesive was applied over the adherend surface and cured to develop an adhesively bonded joint. The adhesive bonding process provides numerous benefits such as uniform distribution of stress over the bonding area, exceptional sealability and relatively reduced stress concentration, as shown in Table 5.1 (Budhe et al. 2017). Thus, the usage of structural adhesive to merge composite structures has gained traction where lightweight, fuel-saving, aerodynamic and vibrational behaviour are essential. However, adhesive bonding has few disadvantages such as that it requires careful surface preparation of the adherend, is affected by different environmental conditions, and is difficult to inspect and repair structures (Banea and Da Silva 2009). Consequently, extensive and systematic investigations were carried out by the researchers on the failure mode and strength of adhesively bonded joints to ensure their safety and reliability (Apalak 2009). Despite these failures, adhesively bonded joints might be subjected to dynamic loading conditions such as a crash, and vibration during their service, resulting in fatigue failure. Therefore, it is essential to know the static and dynamic response of adhesively bonded joints to produce high performance adhesively bonded composite joints. Vibrational behaviour such as fundamental natural frequency, damping factor and mode shapes are essential in their real-time applications (Rajesh and Pitchaimani 2017). Since harmonic and short-term impact loads tend to vibrate the adhesively bonded structures consistently or sporadically, fatigue life becomes a significant consideration. Hence, it is essential to analyse the dynamic response of adhesively bonded joints to make them safe and reliable during their service period. For doing so, several researchers attempted to study the free and forced vibrational behaviour of adhesively bonded joints using experimental and numerical analysis for different joint designs such as single lap joints, double lap joints and tubular joints. The present study aims to give a brief insight into the recent experimental and numerical research work carried out by the researchers on the free and forced vibrational behaviour of adhesively bonded joints.

5.2 DYNAMIC CHARACTERISTICS OF ADHESIVELY BONDED JOINTS

Adhesively bonded structures are used in numerous engineering applications such as marine, aeronautical, civil and robotic structures due to their benefits over conventional joining methods. However, factors such as joint design, material behaviour and composite bonding techniques affect the performance and reliability of adhesively bonded joints during their service. Therefore, it is essential to analyse their dynamic characteristics to develop high-performance adhesive bonded joints. Since harmonic and short-term impact loads tend to vibrate the adhesively bonded structures consistently, that might initiate fatigue damage and severely affect their dynamic characteristics. Dynamic loading of the adhesive joint is a major problem in several engineering applications. Therefore, adhesive joints should be carefully examined to know the dynamic characteristics that make them safe during their service. Dynamic outputs such as natural frequency, modal damping factor and modes shape help to interpret the dynamic behaviour of adhesive joints (Rajesh and Pitchaimani 2016). Consequently, researchers studied the vibrational characteristics such as natural frequency and damping factor of adhesively bonded joints using experimental and analytical methods to understand their dynamic behaviours, as discussed briefly in the following sections.

5.3 FREE AND FORCED VIBRATIONAL BEHAVIOUR OF ADHESIVELY BONDED JOINTS

The recent experimental and numerical works carried out by researchers on the free and forced vibrational behaviour of adhesively bonded joints is discussed in this section. Adhesively bonded aerospace structures are exposed to vibration in different forms and types such as flutter, buffet and noise during their service, causing aircraft failure, imbalance, malfunctions in engines, damage to the airframes and increase in maintenance time. Hence, it crucial to analyse the dynamic response of newly developed adhesively bonded structures to make them safe, reliable and durable during their service. Consequently, researchers examined the dynamic characteristics of adhesive joints. For instance, Gunes and Apalak (2011) studied the influence of geometrical parameters and material thickness (inner and outer tubes) on the free vibrational behaviour of adhesively bonded tubular lap joints using the finite element method. Results disclosed that outer tube thickness plays a major role in enhancing the free vibrational behaviour of tubular joints, whereas overlap length and inner tube thickness have no significant effect on natural frequencies. The study also reported that artificial neural networks and genetic algorithms can be effectively used to predict the natural frequency of adhesive joints. Zeaiter, Challita, and Khalil (2019) examined the influence of mechanical properties and material geometry on the vibrational behaviour of double lap joints using experimental and numerical analysis. Results affirmed that higher adherend thickness and increased overlap length significantly increased the natural frequency. Also, it was reported that adhesive modulus has a slightly enhanced natural frequency, whereas adhesives thickness has a negligible effect. Oyadiji (2001) investigated the effect of adhesive properties on the free vibrational behaviour of adhesively bonded single lap joints using the finite element method. Results indicated that natural frequencies of the beam increase with higher adhesive modulus; meanwhile the Poisson ratio has a slight impact on the natural frequency. Apalak, Ekici, and Yildirim (2006) examined the influence of geometrical parameters such as support length, joint length, adhesive thickness and thickness of stainless steel plates on the free vibrational behaviour of corner joints thickness using finite element analysis and artificial neural networks. Results stated that adhesive thickness has little influence on natural frequencies, whereas all other factors considerably affected the free vibrational behaviour of corner joints. Ingole and Chatterjee (2010) compared the free vibrational behaviour of monolithic beam and adhesively bonded beam. Results indicated that the monolithic beam has a higher natural frequency than the adhesively bonded beam. Also, it was reported that the natural frequency of the adhesively jointed beam was more sensitive to overlap ratio under free-free boundary conditions. Kemiklioğlu and Baba (2019a) experimentally investigated the forced vibrational behaviour of adhesively bonded glass-fibrefibre–reinforced composite under axial impact load. Results revealed that samples taken from the edge of the plate withstood high impact energies (15 J) and more vibration cycles (24×10^5) compared to those taken from the centre of the plate. The study also reported that increasing impact energies along with vibration reduced the joint strength, whereas at lower impact load (10 J), tensile strength improved noticeably. Kemiklioğlu and Okutan Baba (2019b) investigated the effect of cyclic vibration load (50 Hz and 160 N) on the tensile strength of adhesively bonded single lap joints. Results affirmed that a vibrated joint has lower joint strength when compared to non-vibrated lap joints. They also reported that the maximum strength was attained at 12×10^5 vibration cycles. Kaya, Tekelioğlu, and Findik (2004) investigated the dynamic response of adhesively bonded single lap joint using a three-dimensional finite element model. Results depicted that damping reduced the resonance amplitudes.

They also reported that natural frequency decreases with an increase in adhesive layer thickness. Vaziri and Nayeb-Hashemi (2002) examined the influence of voids of the central overlap region on dynamic characteristics of adhesively joined tubular joint under axial load using a shear lag model. Results stated that a void in the overlap region slightly affects the dynamic response of tubular joints. They also reported that natural frequency increases to specific overlap length; further, increasing the overlap length decreases the natural frequency. Dhilipkumar and Rajesh (2021) examined the influence of multiwall carbon nanotube addition on the free vibrational behaviour of glass-fibrefibre–reinforced composite fabricated using the co-cure bonding method. Results portrayed that the addition of 1 wt% carbon nanotubes considerably enhanced the natural frequency of co-cured composite due to higher interfacial interaction between the fibre and matrix. They also reported that the addition of lower wt% carbon nanotubes enhanced the modal damping factor due to higher interaction between fibre and matrix. Therefore, it is vital to consider the effect of nanoparticle addition during the design of adhesively bonded joints to develop high-performance structures. Wang, Li, and Xie (2019) investigated the free vibrational characteristics of single lap joints and two-step lap joints using finite element analysis. Results revealed that factors such as overlap length, adhesive thickness and step height ratio considerably affected the free vibration behaviour of adhesively bonded joints. Salloum, Challita, and Khalil (2017) examined the dynamic response of free vibrational response of double lap joints under fixed-free boundary conditions using an analytical model and finite element analysis. Results affirmed that factors such as fibre volume fraction, bond length and adherend thickness remarkably influence the natural frequency, whereas adhesive thickness does not affect it. Apalak et al. (2014) examined the free vibrational response of adhesively joined double cantilever beams using genetic algorithm and artificial neural networks with finite element analysis. They found that natural frequencies increase with richer material composition. Also, they reported that plate length and thickness remarkably influences the free vibrational behaviour, whereas adhesive thickness has less impact on natural frequency and strain energies. Hence, during joint designing, it is important to consider factors such as material geometry, overlap ratio, adherend stiffness, adhesive material and adhesive thickness to develop a highly effective adhesively bonded joint.

5.4 CONCLUSION

There is a growing interest in the use of adhesively bonded joints in assembling subcomponents of aerospace structures like wing skins, tails, ailerons and spoilers. Adhesive bonding is a suitable method for assembling composite materials although stress concentration initiates at lower external loads. Thus, it reduces the strength of adhesively bonded joints. Aircraft are exposed to different environmental conditions such as humidity, temperature and are also subjected to external loading during their service. Therefore, it is essential to understand the static and dynamic response of adhesively bonded joints. In this current review, several methods to analyse the vibrational behaviour of adhesively bonded joints were discussed briefly. Vibrational characteristics such as natural frequency, damping factor and modal strain energies significantly influence the strength of adhesively bonded joints. The studies also reveal that factors such as fibre volume fraction, bond length, adherend thickness and material composition remarkably influence the natural frequency, whereas adhesive thickness has no influence on it. Furthermore, it has been found that artificial neural networks and genetic algorithms can be effectively used to predict the natural frequency of adhesive joints using finite element analysis.

REFERENCES

M. K. Apalak, "Free Vibration Analysis and Optimal Design of the Adhesively Bonded Composite Single Lap and Tubular Lap Joints," *Composite Materials Technology: Neural Network Applications* 15, no. 6 (2009): 251–90.

M. K. Apalak, R. Ekici, and M. Yildirim, "Optimal Design of an Adhesively-Bonded Corner Joint with Single Support Based on the Free Vibration Analysis," *Journal of Adhesion Science and Technology* 20, no. 13 (2006): 1507–28.

Z. G. Apalak et al., "Free Vibration Analysis of an Adhesively Bonded Functionally Graded Double Containment Cantilever Joint," *Journal of Adhesion Science and Technology* 28, no. 12 (2014): 1117–39.

I. A. Ashcroft, D. J. Hughes, and S. J. Shaw, "Adhesive Bonding of Fibre Reinforced Polymer Composite Materials," *Assembly Automation* 20, no. 2 (2006): 150–61.

M. D. Banea and L. F.M. Da Silva, "Adhesively Bonded Joints in Composite Materials: An Overview," *Proceedings of the Institution of Mechanical Engineers, Part L: Journal of Materials: Design and Applications*, January 1, 2009.

S. Budhe et al., "An Updated Review of Adhesively Bonded Joints in Composite Materials," *International Journal of Adhesion and Adhesives* 72 (2017): 30–42.

T. Dhilipkumar and M. Rajesh, "Effect of Using Multiwall Carbon Nanotube Reinforced Epoxy Adhesive in Enhancing Glass Fibre Reinforced Polymer Composite through Cocure Manufacturing Technique," *Polymer Composites*, no. April (2021): 1–15.

R. Gunes, M. K. Apalak and M. Yildirim (2011) "Free Vibration Analysis of an Adhesively Bonded Functionally Graded Tubular Single Lap Joint,"*The Journal of Adhesion*, 87:9, 902–925.

X. He and S. O. Oyadiji, "Influence of Adhesive Characteristics on the Transverse Free Vibration of Single Lap-Jointed Cantilevered Beams," *Journal of Materials Processing Technology* 119, no. 1–3 (2001): 366–73.

H. Heidary, N. Z. Karimi, and G. Minak, "Investigation on Delamination and Flexural Properties in Drilling of Carbon Nanotube/Polymer Composites," *Composite Structures* 201, no. November 2017 (2018): 112–20.

S. B. Ingole and A. Chatterjee, Vibration analysis of single lap adhesive joint: experimental and analytical investigation "Journal of vibration and control," 17 no. August (2010): 1547–1556.

A. Kaya, M. S. Tekelioğlu, and F. Findik, "Effects of Various Parameters on Dynamic Characteristics in Adhesively Bonded Joints," *Materials Letters* 58, no. 27–28 (2004): 3451–56.

U. Kemiklioğlu and B. O. Baba, "Mechanical Response of Adhesively Bonded Composite Joints Subjected to Vibration Load and Axial Impact," *Composites Part B: Engineering* 176, no. August (2019a).

U. Kemiklioğlu and B. O. Baba, "Vibration Effects on Tensile Properties of Adhesively Bonded Single Lap Joints in Composite Materials," *Polymer Composites* 40, no. 3 (2019b): 1258–1267.

J. Kupski and S. Teixeira de Freitas, "Design of Adhesively Bonded Lap Joints with Laminated CFRP Adherends: Review, Challenges and New Opportunities for Aerospace Structures," *Composite Structures* 268, no. December 2020 (2021): 113923.

M. Rajesh and J. Pitchaimani, "Dynamic Mechanical Analysis and Free Vibration Behavior of Intra-Ply Woven Natural Fibre Hybrid Polymer Composite," *Journal of Reinforced Plastics and Composites* 35, no. 3 (2016): 228–242.

M. Rajesh and J. Pitchaimani, "Experimental Investigation on Buckling and Free Vibration Behavior of Woven Natural Fibre Fabric Composite under Axial Compression," *Composite Structures* 163 (2017): 302–11.

E. Salloum, G. Challita, and K. Khalil, "Parametric Investigation of Free Vibration of Double Lap Composite Joints," *Proceedings of the 7th International Conference on Mechanics and Materials in Design*, Albufeira/Portugal, June (2017):11–15.

A. Vaziri and H. Nayeb-Hashemi, "Dynamic Response of Tubular Joints with an Annular Void Subjected to a Harmonic Torsional Load," *Proceedings of the Institution of Mechanical Engineers, Part K: Journal of Multi-Body Dynamics* 216, no. 4 (2002): 361–70.

S. Wang, Y. Li, and Z. Xie, "Free Vibration Analysis of Adhesively Bonded Lap Joints through Layerwise Finite Element," *Composite Structures* 223, no. January (2019): 10943.

A. Zeaiter, G. Challita, and K. Khalil, "Investigation of Vibration Modes of a Double-Lap Bonded Joint," *SN Applied Sciences* 1, no. 5 (2019): 1–15.

6 Introduction on Repair of Composites

Types of Repair Techniques

Natesh M
V.S.B Engineering College, Karur, India

Soundhar A
Sri Venkateswara College of Engineering (SVCE), Chennai, India

Anita Jessie and Lakshmi Narayanan V
Vellore Institute of Technology, Vellore, India

CONTENTS

DOI: 10.1201/9781003200994-6

6.1   INTRODUCTION

Composite materials are essential for specialised applications such as aerospace, military, general aviation, automotive, surface transport and sports equipment markets. Usage of composite materials naturally led to a demand to develop repair procedures. Several damages (cracks and dents) in composite products are easily analysed through the naked eye. Some of the damages are best analysed through appropriate non-destructive test (NDT) for structural applications (Armstrong et al. 2005). On the other hand, simple tapping on the surface of composite can be helpful to identify the damaged areas. The sound of the damaged areas provides dull response as compared to other areas. The boundary between good and damaged areas can also easily be identified through the difference in sound for repair. For composite damage, inspection should be conducted in the regular maintenance schedules for structural applications (Ganesh and Chawla 2005). Special interest is shown to the areas which are more susceptible to damage. Repairs in aircraft structures are performed as per the aircraft structural repair manual (SRM). Repairs in other applications are carried out as per the original specification as well as mechanical performance of the components. This section presents to offer a general approach to repairs in composite materials for all applications and will determine the laminate and sandwich structures.

6.1.1   SANDWICH STRUCTURES

Separate thin and high-strength skins are bonded in lightweight honeycomb cores. The stiffer panels are made by thickening the core where weight increase is minimum. Sandwich structure is shown in Figure 6.1.

6.1.2   LAMINATE STRUCTURES

Laminate structures are assembled by fibre orientation and polymer matrix. Most of the required mechanical properties are provided by the fibre orientation and matrix, offering environmental performance. Laminate structure is shown in Figure 6.2.

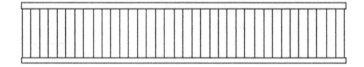

FIGURE 6.1   Sandwich structure.

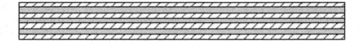

FIGURE 6.2 Laminate structures.

6.2 DAMAGE MECHANISM-COMPOSITE

Due to the anisotropic characteristic of composites, a complex damage mechanism is observed; the properties were directional dependent as well. Numerous parameters (the boundary conditions and layup sequence, as well as the presence of structural flaws during production) can affect the fatigue characteristics of composites (fibre and matrix properties). Plastic deformation of materials also plays a major role when applying the impact energy. In other words, brittle fibre composites (carbon or glass) have low energy absorption capability in comparison with ductile materials (metals). Glass-fibre–reinforced polymers or carbon-fibre–reinforced polymer exhibit a low-impact energy absorption (Naebe et al. 2016). Subsequently, mechanical energy can propagate inside the composite, which can affect barely visible impact damages (BVID). Some of the BVID in composites comprise interfacial debonding, matrix cracking, fibre breakage or delamination. The evolution of damage in composites during applied impact load is shown in Figure 6.3.

In fibre-reinforced plastics (FRP), layers of fibres are kept together by the bonding of the matrix. Therefore, fatigue is started via the method of matrix cracking, subjecting the FRP to damage. The crack then begins to propagate nearer to the fibre, which brings interfacial debonding and additional growth of the crack. Interlaminar crack propagation or delamination is formed after the damage initiation process.

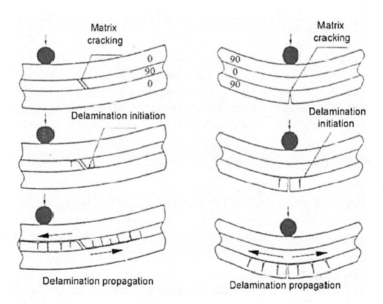

FIGURE 6.3 Various damage mechanisms in composites when they are subjected to impact load.

These cracks or delamination are among the most serious damage mechanisms in polymer-based composites (Kang and Kim 2000). These cracks occur due to their high reduction of compressive strength and the stiffness of the material. It is inferred from literature that interfacial delamination reduces the flexural strength more (up to 46.7%) than that of their non-damaged composites when they are subjected to transverse load. Additional damage will lead to breakage of fibre which will aid in structural failure (Trias et al. 2012). To maintain the performance of the composite products in aerospace structures, safety inspections such as by Federal Aviation Administration (FAA) and European Aviation Safety Agency (EASA) are followed. In the automobile sector, customers will bring their automobiles for inspection when there is any type of impact or corrosion (visual damage). Hence, several types of damages (visible damages) such as matrix cracking, fibre breakage or delamination and interfacial debonding must be examined during the repair.

6.3   KEY STAGES OF COMPOSITE REPAIR

The fundamentals of composite repair remain mostly similar irrespective of how they are made or where they are used.

Techniques are:

- Evaluate to analyse damage (extent and degree);
- Eliminate damaged portion;
- Patch up contaminated portion;
- Fix repair region;
- Complete composite repair; (e.g. change damaged portion and bond or bolt a patch);
- Examine repair for quality assurance (e.g. proper cure, inclusions, delaminations, etc.); and
- Reinstate surface finish.

The choices made in designing a structure of composite such as thickness, curvature and shape, practice of bonded integral stiffeners and cored construction or solid laminate, determine which classes of inspection and repair will be inexpensive and even conceivable, in order to affect the overall cost of the product's life-cycle (Kwak et al. 2019).

6.4   DAMAGE ASSESSMENT

The clearest and most straightforward approach for examining composite damages such as dents and cracks is visual inspection as well as BVID. On other hand, non-visible damages in composites are analysed through non-destructive testing (NDT). Still, there is no particular technique available for assessing all classes of damages.

6.4.1   INFRARED THERMOGRAPHY

Infrared thermography (IRT) is a contactless method which utilises infrared radiation to find any cracks or product damage. IRT is an appropriate method for finding damage in composite products due to a low thermal conductivity in composites. Researchers carried out many works on composites through IRT to obtain impact damages, fatigue damages and delamination (Guillaumat et al. 2004). There are two different approaches (active and passive) in IRT. The passive approach is utilised to study materials with temperatures other than the room temperature. IRT utilises the thermal scan in the composite product without subjecting the composite to external heat. Passive thermography setup is shown in Figure 6.4a. Active thermography (Figure 6.4b) uses a controlled heat input by making thermal variation within the material.

There will be a temperature variation in the specimen's surface because of a variation in the thermal conductivity between the flaws and flawless portion. IRT is utilised for specimen thickness of 2–16 mm. Flaws can show in the surface of thick specimens (>16 mm). Wind blade damage inspections are identified by varying the ambient air temperature between day and night through passive IRT. A small change in temperature was found while analysing the defect in the automobiles (Usamentiaga et al. 2013). This will decrease the application of passive IRT to find major structural discontinuities. In active thermography, pulsed thermography (PT) and pulsed phase thermography are the most frequently utilised thermal stimulating methods to examine composites. IRT has some merits and demerits regarding its application in an inspection. The main merit of IRT is that it permits the detection of damage at just one side of the specimen. Major demerits are the expensive investment in IR devices, accessories and most importantly the requirement of highly skilled persons to examine the flaws (Maier et al. 2014).

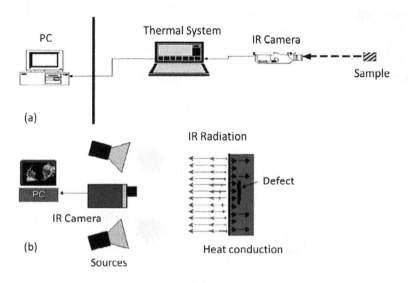

FIGURE 6.4 Setup of (a) active and (b) passive thermography.

6.4.2 ULTRASONIC TESTING

High frequency acoustic energy is utilised in ultrasonic testing (UT) to obtain the defects or flaws in the composite products (D'Orazio et al. 2008). There are two different approaches (transmission and reflection method) utilised in UT. A UT machine comprises of various devices (receiver, pulser, transducer and display devices). An electrical high-voltage pulse will be produced through the receiver, which gives electrical energy to the transducer. Subsequently, the transducer produces a high frequency acoustic energy by using piezoelectric ceramics. This acoustic energy spreads in the form of waves. The pulser and receiver are kept on the same side of the specimen in the reflection method. In the transmission method, the pulser is kept on the exact opposite side from the receiver. The reflected wave signal response time is different coming from the damaged and undamaged portions (Edwards et al. 2006). Fundamental principles of ultrasonic analysis are given in Figure 6.5. UT equipment is used in both contact and contactless basis analysis. From the various research articles, UT for composites is used to analyse the impact damage, cracks and delamination. The merits of UT method are adequacy of single-side access to identify the flaws, the higher depth of penetration to obtain the defects as compared to other NDT techniques, minimal specimen preparation, high accuracy for examining the shape and size of the flaw and instant as well as automated results. The demerits of the UT method are irregular or rough shape, very thin, very small, non-homogeneous samples that are tough to analyse; also, highly skilled persons are needed for UT (Ibrahim et al. 2017).

6.4.3 DIGITAL SHEAROGRAPHY

Digital shearography (DISH) is an optical full-field, laser-based and non-destructive method for finding damages in composites. DISH needs a laser source to light the specimen so that a fleck pattern is produced. The fleck pattern is captured through a video image-shearing camera, which concurrently mixes it with a similar but laterally displaced form of itself. The captured images before and after the loading of the application are refined in a computer and a rectification of the images results in a border or edge pattern. The edge or border pattern will comprise of the displacement derivatives of the exterior of specimen (Groves et al. 2007). The fundamental

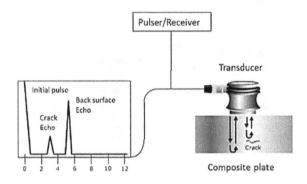

FIGURE 6.5 Schematic diagram of UT technique.

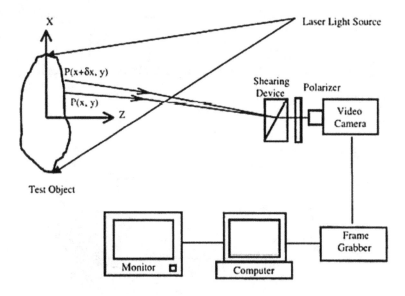

FIGURE 6.6 Schematic diagram of DISH.

principles of DISH are shown in Figure 6.6. DISH is a contactless method to analyse the flaws in honeycomb sandwich composites for the applications of automobiles (Francis et al. 2010). The merits of the DISH method are rapid detection rate and a simpler device which doesn't need any additional scanning devices. However, the flaw detection accuracy is based on the depth of the flaw and its size. One disadvantage is that DISH is not appropriate to find the depth of flaw (Hung 1999).

6.5 TYPICAL DAMAGE

Fibre-reinforced composites are affected due to low velocity impacts as well as occasional high velocity impacts. These composites are generally affected more severely inside than compared to their surface (Heida Jaap and Platenkamp 2011). The damages in the composites are shown in the Figure 6.7–6.9.

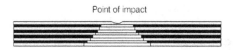

FIGURE 6.7 Delamination following impact on a monolithic laminate.

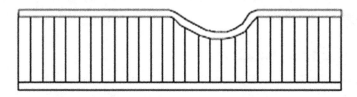

FIGURE 6.8 Dents in sandwich structure.

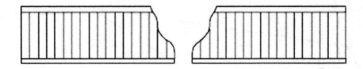

FIGURE 6.9   Puncture damage in a sandwich structure.

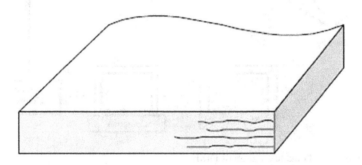

FIGURE 6.10   Laminate splitting.

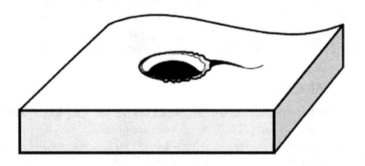

FIGURE 6.11   Bolt hole damage.

6.5.1 LAMINATE SPLITTING

Damage to the composite that does not propagate through the full length of the component is noticed. The impact on mechanical behaviour is based on the length of split with respect to the product thickness. Laminate splitting is shown in Figure 6.10.

6.5.2 BOLT HOLE DAMAGE

In composite products, damage can occur to the hole which affects the laminate splitting or harm to the upper plies. Bolt hole damage is shown in Figure 6.11.

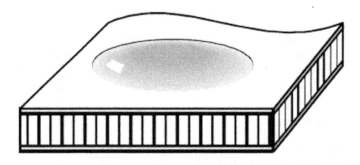

FIGURE 6.12 Heat Damage.

6.5.3 Heat Damage

Heat damage occurs in composites by separating the surface plies, which occurs as a local fracture. The heat damage affects mechanical performance based on the thickness of the component. Heat damage is shown in Figure 6.12.

6.6 REPAIR OPTIONS

When servicing the composite product, repairs must be carried out in three levels to eliminate the damage.

6.6.1 Cosmetic Repair

In the case that analysis has evaluated that the damage has not infected the structural integrity of the part, a cosmetic repair is performed to save and improve the appearance of the surface. Use of reinforcement is not involved in this method.

6.6.2 Temporary or Interim Repairs

In lot of cases during service, the tiny region of damage will not affect the mechanical performance of the product. However, the unrepaired tiny damage may be susceptible to additional propagation through environmental conditions and fatigue. Repairs must be performed using a simple patch as well as minimum preparation to save the product until it can be cut off for a suitable repair. These temporary repairs must be applied during a regular inspection.

6.6.3 Structural Repair

Several damages (debonding or delamination and fibre fracture) in the composites need to be addressed by changing the damaged reinforcements and core in sandwich components to regain its original mechanical properties. The bonded-on repair seen with a break in the original plies needs to be changed with new plies to repair the structure.

6.7 LAMINATES AND SANDWICH PANELS

A significant aim of a structural repair is to completely support given loads and transfer given stresses across the repaired region. To carry out this repair, plies in composites must be overlapped and bonded in the laminate. Following are three different approaches used to perform the repair (Takeda et al. 2007; Amaro et al. 2013).

6.7.1 PATCH REPAIR

In patch repair, filler plies are used to recreate the thickness of the original laminate and the repair constituents are joined to the surface of the laminate.

Advantages

- Rapid and easy to perform
- Needs minimum preparation

Disadvantages

- A rectified laminate is heavier and thicker as compared to the original
- For attaining good adhesion, very cautious surface preparation is required

6.7.2 TAPER SANDED OR SCARF REPAIR

In the scarf or taper sanded repair, the region around the hole in a composite is sanded to show a section of each ply in the laminate. In some cases, a single ply is used to create a flat surface in the tapered material (30–60:1).

Advantages

- Repair is only slightly thicker as compared to the original
- Each repair ply joins the ply that it is rectifying and offers a strong and straight load path
- Good bonds can be obtained on the newly visible surfaces

Disadvantages

- Time-consuming process
- Highly trained person required and difficult to obtain

6.7.3 STEP SANDED REPAIR

The laminate is sanded down with the intention that a flat band of every layer is shown, making a stepped finish. These steps have been with the thickness of 25–50 mm in a layer.

Advantages

- Similar to taper sanded repair

Disadvantages

- Very difficult to carry out

6.8 TYPICAL LAMINATE REPAIRS

Laminate repairs such as patch repair, taper sanded repair and step sanded repair are shown in Table 6.1. Note: If the part has been in regular service, it must be dried to eliminate any wetness to achieve the best repair (Takeda et al. 2007).

6.9 TYPICAL SANDWICH PANEL REPAIRS

Sandwich panel repairs such as patch repair, alternate patch repair, taper sanded repair and step sanded repair are shown in Table 6.2. Note: If the part has been in regular service, it must be dried to eliminate any wetness to achieve the best repair (Amaro et al. 2013).

TABLE 6.1
Types of Laminate Repair Techniques

Sl.No	Types of laminate repairs
1	Patch repair

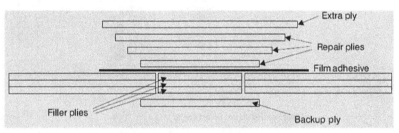

2	Taper sanded repair (Scarf repair)

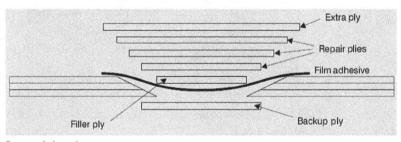

3	Step sanded repair

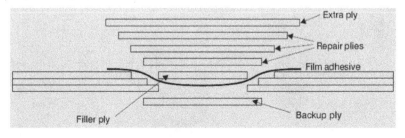

TABLE 6.2

Types of Sandwich Panel Repairs

Sl.No	Types of sandwich panel repairs
1	Patch repair

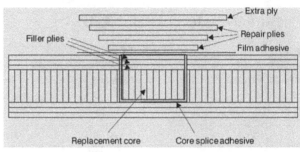

| 2 | Alternative patch repair |

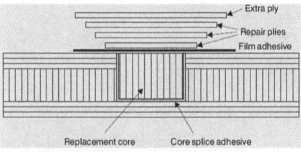

| 3 | Taper sanded repair (scarf repair) |

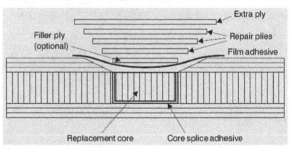

| 4 | Step sanded repair |

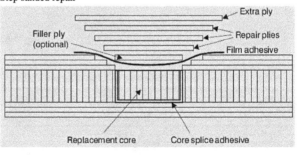

6.10 REPAIR USING PRE-CURED DOUBLER

Precured doublers are used to repair several composite structures. A sheet of composite is produced from the epoxy (matrix) and fibre (reinforcement) that has been cured using pressure and heat, known as pre-cured doubler. For performing the repair, a patch of precured doubler is used to bond the surface of the product using adhesive films or wet lay-up resins. The edges in precured doubler are chamfered to increase peel-off strength. The demerit of this method is that trapped air between the doubler and the layer of product that makes this a weak method. To remove the air gap, a layer of scrim cloth is used under the adhesive film (Roach and Rackow 2018).

6.11 REPAIR PROCESS

Autoclave and vacuum bag processing are the two significant techniques for the repair of parts from prepreg. This technique is determined by the cost, type of part being produced and quality.

6.11.1 VACUUM BAG PROCESSING

This method is appropriate for parts with thin sections and huge sandwich structures. This method involves the placing and sealing of a flexible bag on a composite lay-up and removing all the air from under the bag. The evacuation of air forces the bag down onto the lay-up with the pressure of 1 bar. The assembly (with vacuum still given) is kept inside an oven with good air flow and the composite is made after a moderately short duration (Rider et al. 2011).

6.11.2 AUTOCLAVE PROCESSING

This technique is utilised for the repair of advanced structural applications. This technique uses a similar vacuum bag (Figure 6.13) but the oven is replaced by an autoclave. The autoclave is a pressure vessel that offers various curing settings for the composite. This technique involves pressure, vacuum, cure temperature and heat-up rate that can be controlled. Products with high thickness and complex shapes are repaired using high pressure in the autoclave process (Robson et al. 1992).

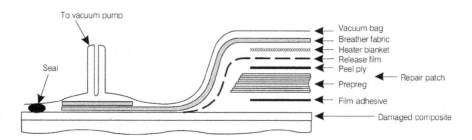

FIGURE 6.13 Detail of vacuum bag lay-up.

6.12 SUMMARY

Composite structures are used in many significant applications such as automobiles and aerospace. Usage of composite structures eventually led to develop repair procedures encompassing various parameters. Numerous parameters (the boundary conditions and lay-up sequence, as well as the presence of a structural flaw during production) can affect the fatigue characteristics of composites (fibre and matrix properties). Additionally, plastic deformation of materials plays a major role while applying impact energy. Impact energy can propagate inside the composite, which is the reason for higher damages in composite products. Matrix cracking, fibre breakage or delamination, and interfacial debonding are major damages affecting the composite structures. Several damages can be identified with the naked eye. Some internal damages can be only found through the NDT methods (infrared thermography (IRT), ultrasonic testing (UT) and digital shearography (DISH)). IRT is utilised for specimen thickness of 2–16 mm as flaws can show in the surface of thick specimens (>16 mm). UT is used for identifying damage in thick (<16 mm) and regular-shaped composite structures. DISH is employed to find the damages in thin specimens. Various damages are rectified through different repair techniques. The repair techniques (cosmetic repair, temporary or interim repairs and structural repair) are based on the damage in the surface of the composite or in-depth damage. Autoclave and vacuum bag processing are the two significant techniques for the repair of parts from prepreg. This technique is determined by the cost, type of part being produced and quality.

REFERENCES

Amaro AM, Reis PNB, De Moura MFSF, and Neto MA. "Influence of multi-impacts on GFRP composites laminates." Composites Part B: Engineering 52 (2013): 93–99.

Armstrong K, Cole W, and Bevan G. Care and repair of advanced composites. SAE, 2005.

D'Orazio T, Leo M, Distante A, Guaragnella C, Pianese V, Cavaccini G. Automatic ultrasonic inspection for internal defect detection in composite materials. NDT&E International 2008;41:145–54.

Edwards RS, Dixon S, Jian X. Characterisation of defects in the railhead using ultrasonic surface waves. NDT&E International 2006;39:468–75.

Francis D, Tatam RP, Groves RM. Shearography technology and applications: a review. Measurement Science and Technology 2010;21:102001.

Ganesh VV, Chawla N. Effect of particle orientation anisotropy on the tensile behavior of metal matrix composites: experiments and microstructure-based simulation. Materials Science and Engineering 2005;391:342–53.

Groves RM, Chehura E, Li W, Staines SE, James SW, Tatam RP. Surface strain measurement: a comparison of speckle shearing interferometry and optical fibre Bragg gratings with resistance foil strain gauges. Measurement Science and Technology 2007;18:1175–84.

Guillaumat L, Batsale JC, Mourand D. Real time infra-red image processing for the detection of delamination in composite plates. Composite Applied Science Manufacturing 2004;35:939–44.

Heida Jaap H, Platenkamp DJ. Evaluation of non-destructive inspection methods for composite aerospace structures. International Workshop of NDT experts, Prague, vol. 6. 2011. p. 1–12.

Hung YY. Applications of digital shearography for testing of composite structures. Composites Part B: Engineering 1999;30:765–73.

Ibrahim ME, Smith RA, Wang CH. Ultrasonic detection and sizing of compressed cracks in glass- and carbon-fibre reinforced plastic composites. NDT&E International 2017;92:111–21.

Kang TJ, Kim C. Impact energy absorption mechanism of largely deformable composites with different reinforcing structures. Fibres and Polymers 2000;1:45–54.

Kwak BS, Lee GE, Kang GS, and Kweon JH. "An investigation of repair methods for delaminated composite laminate under flexural load." Composite Structures 215 (2019): 249–257.

Maier A, Schmidt R, Oswald-Tranta B, Schledjewski R. Non-destructive thermography analysis of impact damage on large-scale CFRP automotive parts. Materials 2014;7:413–29.

Naebe M, Abolhasani MM, Khayyam H, Amini A, Fox B. Crack damage in polymers and composites: A review. Polymer Reviews 2016;56:31–69.

Rider AN, Baker AA, Wang CH, and Smith G. "An enhanced vacuum cure technique for on-aircraft repair of carbon-bismaleimide composites." Applied Composite Materials 18, no. 3 (2011): 231–251.

Roach D, and Rackow K. "Development and validation of bonded composite doubler repairs for commercial aircraft." In Aircraft sustainment and repair, pp. 545–743. Butterworth-Heinemann, 2018.

Robson JE, Matthews FL, and Kinloch AJ. "The strength of composite repair patches: a laminate analysis approach." Journal of Reinforced Plastics and Composites 11, no. 7 (1992): 729–742.

Takeda N, Minakuchi S, and Okabe Y. "Smart composite sandwich structures for future aerospace application-Damage detection and suppression-: A review." Journal of Solid Mechanics and Materials Engineering 1, no. 1 (2007): 3–17.

Usamentiaga R, Venegas P, Guerediaga J, Vega L, López I. Automatic detection of impact damage in carbon fibre composites using active thermography. Infrared Physics & Technology 2013;58:36–46.

7 Adhesively Bonded Composite Joints in Aerospace Application
An Overview

Thulasidhas Dhilipkumar and Murugan Rajesh
Vellore Institute of Technology, Vellore, India

CONTENTS

7.1 INTRODUCTION

Advanced fibre-reinforced composites are replacing metallic materials in aerospace, marine, automobile and robotic applications. The high strength and exceptional stiffness of these materials facilitate lightweight and high-performance structures (Mostafa et al. 2017). Classically, these structures are combined using mechanical fastening due to their easy assembling and dismantling. They also need significantly less surface preparation, and it's easy to inspect the quality of the joint. However, mechanical joints are held together with bolts or rivets, which necessitate the drilling of holes (Heidary, Karimi, and Minak 2018; Karimi et al. 2013). Drilling holes in composite laminates causes fibre damage, resin degradation, delamination and stress concentration around the holes. These damages can initiate fatigue crack and reduce the fatigue properties of the composite structure (Chowdhury et al. 2016; Mariam et al. 2018). Therefore, the insertion of fasteners such as bolts, rivets and screws severely affects the benefit of lightweight fibre-reinforced composites (Hu et al. 2018). It also poses complications in joining different material surfaces (Li et al. 2020).

Researchers established adhesive bonding techniques to overcome the disadvantages of mechanical joining (Banea and Da Silva 2009). Adhesive bonding is the suitable joining method in which adhesive is applied between the adherend surface, then cured to produce an adhesive joint. The main advantage of the adhesively bonded joint is that drilling of holes and use of fasteners are not needed, and it ensures relatively lightweight structures. Therefore, the use of structural adhesive to join composite parts has extensively increased in aerospace, submerged vehicle and automotive industries owing to their high strength, flexibility and uniform stress distribution (Banea et al. 2018). The Boeing 787 Dreamliner reduced nearly 50% of its final weight and achieved 20% less fuel consumption due to the efficient use of composite in its structure using the adhesive bonding method (Giurgiutiu 2016). Furthermore, the adhesive bonding method is used to repair damaged composite structures, allowing these repairs to be completed without causing further damages to substrates (Olajide, Kandare, and Khatibi 2017).

Compared to mechanical fasteners such as bolts, screws and rivets, adhesive bonding offers a continuous and larger assembly area, which considerably lowers the stress concentration. However, stress concentrations still exist in adhesive and substrates due to the intrinsic discontinuity of materials at the bonding region. The researchers have found that higher stress was developed at the ends of the bonding area. The significant stresses present in the adhesive and substrates are classed as shear stress and peel stress. Peel stress has an enormous impact on joint strength in adhesive joints with composite substrates because it directly forces the composite matrix in a non-reinforced direction (Gholami, Khoramishad, and da Silva 2020). Generally, the load-carrying potential of the composite structure is governed by its weakest parts. The relationship among substrates, adhesive and substrates interface plays a significant role in the behaviour of the adhesive-bonded joint. The development of different surface preparation methods – for instance, peel ply, abrasion, media blasting and plasma treatment – ensures better interfacial bonding between adhesive and substrates (Guo et al. 2021).

However, bonded connections of composite materials are highly prone to delamination, the early failure mode in which the composites break, because of significant peel stress that takes place in the joints due to the lower transverse rupture strength of the composites (Neto, Campilho, and Da Silva 2012). Kim et al. (2006) predicted the failure of composite lap joint using numerical analysis. Results depicted that the highest joint strength was accomplished when the adhesive and delamination failure happened simultaneously. Panigrahi and Pradhan (2007) analysed the delamination damages in an adhesively bonded fibre-reinforced composite laminate. Results exposed that delamination in adherend affects the joint stiffness and increases stress concentration. Therefore, the delamination-prompted damages may reduce resistance against shear loading. It is one of the critical failure modes in FRP composite joints. Novel techniques are required to prevent early initiation of delamination and enhance the load-carrying ability of adhesively bonded joints. Usually, fibre-reinforced composite has lower transverse strength because fibres in a laminate's plane do not offer reinforcement throughout the thickness; the composite relies entirely on the weak matrix to transmit loads in that direction (Wisnom 2012). Presently, the resins used in advanced composite materials are brittle, with lower toughness than commercially available adhesives (Jojibabu, Zhang, and Prusty 2020). Nowadays, the need for high strength and reliable joints are indispensable in weight-sensitive applications (Dhilipkumar and Rajesh 2021). Therefore, substantial and systematic research on the failure and strength of adhesively bonded joints are essential to make safe and reliable designs. In the current study, factors influencing the performance of the joint such as composite joint technique, surface preparation, geometrical factors (bond line thickness, joint design), material parameters (substrate and adhesive properties) and environmental conditions are discussed elaborately.

7.2 FACTORS INFLUENCING THE PERFORMANCE OF ADHESIVELY BONDED JOINTS

The performance of adhesively bonded joints relies on several factors, for example, composite joint technique, surface preparation, geometrical factors (bond line thickness, joint design), material parameters (substrate and adhesive properties) and environmental conditions. All these factors must be taken into consideration to produce high-performance joints. These will be deliberated in the following subchapters.

7.2.1 COMPOSITE BONDING PROCESS

The bonding technique is one factor to be taken into consideration while preparing a bonded joint. The composite laminate can either be a secondary bonded or co-bonded joint in most applications to replace the mechanical fastener (Shin et al. 2003). The schematic diagram of different composite bonding techniques is shown in Figure 7.1. The co-cure or co-bonding technique is frequently chosen over the secondary bonding method since it reduces the number of components, assembly stress, processing time and curing cycle (Shin and Lee 2000). Hence, it is widely used for repairing composites. On the other hand, the secondary bonding method is better suited for massive and complicated structures. The advantages and disadvantages associated with each of these manufacturing processes are listed in Table 7.1.

In AV-8B harrier aircraft (Watson 1980), the wings were assembled using the co-cure manufacture technique, and it almost excluded 60% of fasteners. Kageyama and Yoshida (2000) fabricated co-cured wings for XF-2 fighter aircraft and reduced 250 kg of its weight compared to conventional structures. Stevenson (2015) examined the flexural response of composite panels manufactured using secondary bonding and

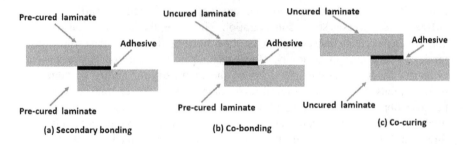

FIGURE 7.1   Schematic diagram of the composite bonding process.

TABLE 7.1
Advantages and Disadvantages of Composite Bonding Techniques

S.NO	Bonding Process	Advantages	Disadvantages
1	Secondary bonding	• Lower stress concentration. • Uniform stress transfer.	• Requires surface preparation. • Requires additional curing time.
2	Co-bonding	• The best technique for joining sandwich and corrugated structures. • Suitable for multi-material bonding.	• Selection of adherend is quite challenging. • Requires surface preparation.
3	Co-curing	• Reduces extra processing steps and curing time. • Simpler and cost-efficient technique.	• Not suitable for large scale and complex assemblies. • Moisture existing in the prepreg may weaken the adhesive layer.

co-curing technique. Results asserted that co-cured composite has 11% higher flexural strength than the secondary bonded composite panel. Ye et al. (2018) analysed the post-buckling behaviour of a T-stiffened panel using co-curing, co-bonding and secondary bonding techniques. Results showed that the manufacturing method has a higher impact on the buckling behaviour of the stiffened panel under axial compressive load. Moroni et al. (2019) examined the tensile properties of composite joints fabricated using the co-cure and co-bond techniques under mode I and mode II loading conditions. Results showed that the co-cured joint has lower fracture strength than the co-bonded composite joint in both loading conditions. Song et al. (2010) studied the effect of the manufacturing process on the failure behaviour of single-composite lap joints fabricated using co-curing with adhesive and without adhesive, secondary and co-bonding. Results revealed that co-cured composite joint without adhesive and secondary bonded composite joint has higher shear strength.

Kim, Choi, and Kweon (2010) investigated the post-buckling behaviour and strength of stiffened composite panels fabricated using different manufacturing techniques. Results revealed that separation failure happened in a single location for co-cured composite panels at higher load. In contrast, in the secondary bonded panel, the separation failure happened in many locations at lower load. It is affirmed that co-cured composite panel has superior strength and post-buckling properties compared to the secondary bonded composite panel. Akin (2018) manufactured a three-beam cell box using the conventional technique, co-curing and secondary bonding. Results demonstrated that the co-cured cell box endured a 95% higher transverse load during the bending test when compared to the secondary bonded cell box. The study revealed that energy and time could be saved by 57% and 25%, respectively, through the co-curing technique. Mohan, Ivanković, and Murphy (2015) analysed the effect of co-curing and secondary bonding manufacturing under mode I and mixed load conditions. The result stated that the co-cure joint has lower joint strength due to moisture in prepreg released during the curing process, which deteriorates the interface in the adhesive layer. From the results, it must be noted that moisture content, curing temperature and adhesive materials must be considered during the preparation of composite joints. For repairing composite structures, selecting an appropriate joining method is critical, as the damaged structure may contain moisture in it. Multimaterial bonding has been widely used to get the combined benefits of metals and composite materials. For instance, the composite material has a lightweight and more structural efficiency than metals, whereas metals have exceptional load-carrying ability and failure predictableness compared to composite materials. Therefore, to utilise the combined benefits allied with metals and composites, multi-material joints between both types of materials are increasingly developed (Lißner et al. 2020). Many researchers have stated that the multi-material bonding technique improved the strength of bonded structures. Using analytical and experimental techniques, Rudawska (2010) analysed the shear strength of lap joints fabricated using different substrate materials such as titanium, aluminium and aramid-epoxy composite. Results indicated that aluminium-aluminium adhesively bonded joints showed higher strength than other combinations. Seong et al. (2008) compared the shear strength of metal-to-metal joints with composite-to-metal joints. Results revealed that metal-to-metal joints have higher strength than composite-to-metal joints. Avendaño et al. (2016) examined the joint strength of dissimilar single lap joints fabricated using the adhesive joining technique under the quasi-equilibrium process and impact loading as a temperature-dependent function. Results disclosed that joint strength was reduced at increased temperature.

7.2.2 SURFACE PREPARATION

The surface preparation of adherend is a critical factor that directly influences the mechanical interconnectivity between adhesive and adherend. Joint strength can be enhanced by ensuring that the surface of the adherend has good wettability and is free from contaminants such as lubricants, rust, dust, etc. Since the adherend surface can be prepared using different physical and chemical treatment methods, the selection of appropriate methods is essential to develop high-strength joints.

Before adhesive bonding, a clean and preferably active surface must be provided through primary and minimal surface pretreatment. Same-surface pretreatment methods have been used by researchers, such as peel ply technique, mechanical polishing, sandblasting and anodizing. In the peel ply technique, a peel ply fabric is utilised as a detachable layer peeled off to change the surface for adhesive joining. It also protects the adherend surface from contamination. Kanerva and Saarela (2013) reported that reviewing the correlation between peel ply surface treatment and bond strength is complicated. In an adhesively bonded composite joint, the adhesion and its performance directly relied on the type of adhesive used. Budhe et al. (2017) analysed the surface of peel ply treated composite adherend using scanning electron microscopy. The surface morphology results indicated that matrix material interacts with peel ply fabric and leaves residues of peel ply on the adherend surface. Therefore, careful selection of peel ply and its removal is essential to produce high-performance composite joints. Boutar et al. (2016) stated that mechanical polishing is a simple and effective method to increase surface roughness. Various grades of emery sheets can be used to modify the substrate surface to varied roughness levels (Shokrian, Shelesh-Nezhad, and Najjar 2018). Sandblasting is another effective physical surface treatment procedure that eliminates machining marks and alters the adherend surface and its structure. Bonpain and Stommel (2018) prepared different aluminium alloy samples using the sandblasting method, with various surface roughnesses from 0.15 μm to 18 μm. Results indicated that when roughness increases, lap shear strength gradually approaches a peak value initially, indicating an adhesive failure mode, then declines to a minimum value, indicating a mixed failure mode.

Anodising is an effective technique (Correia, Anes, and Reis 2017) to enhance the lap shear strength of aluminium alloys and provides higher resistance against corrosion. Phosphoric acid anodising is widely applied in aerospace applications to improve the assembly area by developing an oxide layer with nano-pores on the adherend surface. Several studies concluded that phosphoric anodising was an appropriate technique for increasing bonding strength. Also, sulfuric acid anodising can develop a fine oxide layer on the adherend surface. It is employed in civil and automotive applications. Also, it is comfortable for high-humidity locations. Meanwhile, chromic acid anodising is prohibited due to the carcinogenesis characteristics of hexavalent chromium, and it also has the potential to contaminate the environment (Capelossi et al. 2014). Furthermore, techniques like laser and plasma etching can drastically alter the adherend surface's physical and chemical characteristics that will increase joint strength.

7.3 GEOMETRICAL FACTORS

7.3.1 ADHESIVE THICKNESS

Adhesive thickness is one of the major design factors in adhesively bonded joints as it significantly affects joint strength (Silva et al. 2006). Therefore, to understand the influence of adhesive thickness on the performance of adhesive joints, researchers carried out many investigations under different loading conditions using experimental and numerical analysis. Generally, in single-lap joints, the shear strength decreases as the bond line thickness increases due to voids, cracks and high interfacial stress. Carbas et al. (2021) analysed the shear properties of the butt joint under torsional loading. Results depicted that increasing bond line thickness reduced the shear properties during the torsion test. However, the elasticity of the material remains constant. Also, the study suggested that torsional testing should be carried out for adhesive joints with an adhesive thickness of 0.5 mm because of its accurate results without altering other mechanical properties.

Naito, Onta, and Kogo (2012) analysed the influence of adhesive thickness on shear and tensile properties of butt and single lap joints prepared using polyimide adhesive. Results revealed that increasing bond line thickness decreased the strength of butt joints. Meanwhile, bond line thickness did not affect the shear properties of single lap joints. Pascoe et al. (2020) investigated the effect of bond line thickness on fatigue crack propagation of (FM94) epoxy adhesive film using experimental and numerical analysis. Results indicated that increased adhesive thickness promoted early crack growth. Also, it reported that energy needed for per unit crack growth does not rely on the bond line thickness, while the energy existing for crack growth depends on bond line thickness. Furthermore, there is no overall relationship between strength and adhesive thickness. This mixed behaviour can be attributable to several factors such as the type of adhesive material, loading conditions, substrate behaviour and joint geometry, which alters the performance of adhesive joints as their bond line thickness is varied.

7.3.2 Joint Design

The joint design plays a significant role in the strength of adhesively bonded joints. The joint design should provide uniform stress distribution across the bonding area, and any design causing stress concentrations, peel stress and high interfacial stress should be eliminated because it promotes early failure initiation. The designer has access to a wide range of joint configurations such as single- and double-lap joints, stepped joints, butt-joint joints, and scarf joints, as shown in Figure 7.2. Generally, single-lap joints are preferred over other joints due to their simple design and efficiency.

Stress concentration mainly occurs at the ends of the overlap region, severely affecting joint strength and reducing the resistance against shear loading. Researchers developed different joint geometries to reduce stress concentration. Kishore and Prasad (2012) fabricated a flat joggle joint using the hand lay-up technique. Results revealed that modified design aided to avoid bending effects and increased the joint strength by 90% compared to unmodified single-lap joints. Taib et al. (2006) reported that the joggle joint increased the strength of steel parts compared to the conventional joint.

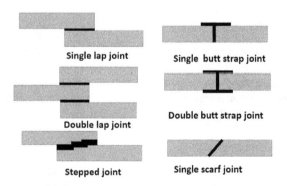

FIGURE 7.2 Joint Configurations.

Zheng et al. (2020) analysed the influence of spew fillet on single-lap joints fabricated using carbon fibre and aluminium alloy dipped using distilled water. Results revealed that spew fillets increased the shear strength of the lap joint. Zielecki et al. (2017) investigated the fatigue properties of adhesively joined steel parts using different size chamfer and fillets. Results affirmed that joint strength could be increased using chamfers and fillets. Lucas and Adams (2007) reported that internal taper and fillet increases the bond strength by reducing peel stress.

Researchers developed a wavy type lap joint to enhance the shear strength of the joint. For instance, Ávila and Bueno (2004) investigated the shear strength of glass-fibre–reinforced epoxy wavy lap joint. Results depicted a 41% improvement in shear strength compared to unmodified single lap joints due to the development of compressive stress near the overlap region's tip. In general, the wavy joint seems to be an intriguing structural optimization approach for converting out of plane tensile load (peel stress) into compressive stresses. Although the overlap zone's intricate shape might be a drawback in an aerospace application like fuselages, which require circumferential joints in their panels, and where aerodynamics features are critical.

The scarf and stepped joint were developed by researchers for repairing composites (Xiaoquan et al. 2013). The stepped joint consist of a single overlap in which the substrate loses half of its thickness for a specific length of the overlap region. Thus, it may reduce peel stress. Wu et al. (2018) compared the damage tolerance of scarf and stepped joints. Results exposed that stepped joint has better damage tolerance than scarf joints. The finger joints are also used to enhance the joint strength of the composite due to their smooth stress transfer and reduced peel stress. Finger joints are also called tongue and groove joints, widely used in wood industries where the slots are created using rotational milling. The finger joint is suited to substrates having a minimum thickness of 5 mm (Lopes et al. 2015). Kupski et al. (2019) stated that the stacking sequence of laminates affects the failure mode and load-bearing behaviour of the composite joints. Ozel et al. (2014) reported that varying stacking sequences can increase joint strength by 120%. Thus, the proper selection of stacking sequence can lower the stress concentration in composite joints. Hence, it is critical to have the correct stacking sequence since it directly corresponds to the adherend thickness. Finally, it was concluded that factors such as adhesive thickness, substrate thickness, stacking sequence, chamfer, fillet, etc. significantly affected the strength of adhesively bonded joints.

7.4 MATERIALS

The selection of appropriate material (adhesive and substrate material) is vital because it has an overall impact on the performance of adhesively bonded joints. Therefore, to understand the influence of failure mode, adhesive and substrate in composite bonded joints, researchers conducted several experimental and numerical studies, which are detailed in the following subsections.

7.4.1 ADHESIVE MATERIAL

It is essential to differentiate adhesive strength and bond strength because bond strength may not get enhanced if a more rigid adhesive is used. A more challenging and stiffer adhesive will endure high stress, but its rigidity will significantly raise the stress concentration at the overlapping end. The flexible adhesive can provide uniform distribution of stress across the bonding area, but it will lower the stress-bearing capacity of the joint (Teixeira de Freitas and Sinke 2017). To overcome these bottlenecks, extensive studies were conducted by the researchers by changing material properties of adhesive at the overlap region, using different adhesive material at the edges, using mixed adhesives at the middle of the overlap region or using physically modified adhesive at the overlap area (Kupski and de Freitas 2021).

7.4.1.1 Mixed Adhesives

Mixed adhesives were developed to get the combined advantage of more robust and flexible adhesives. At first, mixed adhesives were used in the metallic joint to lower the stress concentration in the adhesive material. (Temiz 2006). Machado et al. (2018) examined the shear strength of carbon-fibre–reinforced substrates prepared using single and mixed adhesives. Results revealed that joints fabricated using mixed adhesive have 64% higher shear strength than single adhesive joints. da Silva and Lopes (2009) analysed the shear properties of composite joints fabricated using brittle adhesive in the middle part of the overlap area; three different ductile adhesives were used in overlap tip. The results stated that the hybrid adhesive approach enhanced the lap shear strength by 221% and 212% compared to brittle and ductile adhesive materials. Öz and Özer (2017) used two different ductile adhesives at the end of the overlap tip and brittle adhesive material in the centre. The experiment's results depicted that mixed adhesive joints increased the joint strength by 40%. However, the separation of different adhesives was complicated in the mixed adhesive approach. Adhesive film can manage the process because of its dimensional constancy, but finding suitable adhesives in film shape is complicated.

7.4.1.2 Physical Alteration of Adhesive Material

The adhesive can be subjected to local physical change by reinforcing rubber particles or using different nanofillers such as carbon nanotube, silica, alumina, nano clay and graphene. For instance, Sancaktar and Kumar (2000) studied the effect of rubber-particles–reinforced adhesive at the end of the overlap area. Results revealed that the incorporation of rubber particles enhanced the joint strength due to the compelling presence of rubber particles at the crack, reduced the local stress concentration at the overlap tip and increased the resistance against loading. Bagheri and Pearson (1996) reported that there is a specific particle size that is ideal for attaining maximal adhesive toughness. The uniform dispersion of the nanoparticles with a fixed weight fraction is essential to increase the joint strength. However, the toughening effect is insignificant if the rubber particles are small. Also, homogenous dispersion of rubber particles into the adhesive is a complex process.

7.4.1.3 Smart Adhesive

Innovative adhesives such as self-healing, dis-bond adhesive materials have been used presently due to their potential to self-heal the damage and increase the reliability of the structure. It is also an economical process when compared to the repair of damaged structures. Reddy et al. (2020) found self-healing adhesive can be made using cyclopentadienyl and grub catalyst through the microcapsule encapsulated self-healing method. Shanmugam et al. (2019) found that the incorporation of microcapsules improved the fracture toughness of the composite. Self-healing (Jin et al. 2012) can also be accomplished in thermoset epoxy resin using twin microcapsule epoxy amine interaction, although the self-healing technique in adhesive joints is still in the initial stage ofresearch.

7.4.2 ADHEREND MATERIAL

The selection of suitable adherend material is vital since different materials act differently and influence the performance of the joints. The essential parameters are modulus and strength of the adherend material. The adherend material with higher modulus will have lower deformation at overlap ends, where the stress transfer occurs. Meanwhile, adherend material with lower modulus develops differential straining in the adhesive material. Generally, adherend material can be classified into two types: metal and composites. In the case of the metal adherend, yielding of adherend material can promote early failure in the joint.

On the other hand, the lower transverse strength associated with composite adherend tends to fail in interlaminar mode due to higher peel stress at the overlap tip. A considerable difference in joint strength was noticed with different substrate materials during the same loading conditions (Ghumatkar, Sekhar, and Budhe 2017). Furthermore, the multi-material joining technique has been used recently due to the combined advantages of metal and composites. However, the selection of suitable adhesive based on adherend material is essential to develop high-performance joints.

7.5   ENVIRONMENTAL CONDITIONS

Environmental conditions such as temperature and humidity directly influence the performance and durability of the adhesive joint during its service. Ishai (1978) investigated the mechanical properties of glass composite in wet and dry conditions. Results revealed that the wet condition environment increased the tensile strength of the composite compared to the dry condition environment. Compression test results (Walker 2004) indicated that high temperature (177°C) decreased the strength, whereas lower temperature (−129°C) increased the compressive strength. Parker (1986) found that a hot-air environment decreasedjoint strength. Ashcroft, Hughes, and Shaw (2006) investigated the effect of temperature and humidity on the shear strength of composite joints. Results showed that matrix material can affect the joint strength of the composite. Park et al. (2010) compared three fabricating techniques such as co-curing, co-bonding and secondary bonding processes in different environmental conditions. Results stated that co-cure joints have the highest shear strength in all conditions. Hence, it is essential to consider these factors during the design and repair of the adhesively bonded joints.

7.6   CONCLUSION

There is a growing trend in using lesser weight fibre-reinforced polymer matrix composite in several engineering applications. The adhesive bonding technique is appropriate for assembling composite structures even though higher peel stress and lower transverse strength of composites reduce the strength of the joints. The bonding method preferred is typically determined by the adherend to be joined, nature of service and field of application. Furthermore, the selection of suitable bonding techniques while repairing structures is important since damaged material may contain moisture. There is no overall relationship between strength and geometric factors (adhesive thickness, joint design). This mixed behaviour can be attributable to several factors such as the type of adhesive material, loading conditions and substrate behaviour, which alters the performance of adhesive joints. Hence, it is vital to consider these factors during the optimization of geometrical parameters to improve the strength of adhesive joints. The researchers continuously develop mixed adhesive, physically toughened adhesive and innovative adhesive to enhance the performance of adhesive joints. For instance, intelligent self-healing adhesive has been used recently due to its potential to heal the damaged structure. Adhesive joints are exposed to environmental conditions such as temperature and humidity that considerably influence the performance and durability of the structure during their service. Hence, it is essential to design an adhesive joint by considering all these factors.

REFERENCES

Akın, M. "*Co-Cured Manufacturing of Advanced Composite Materials Using Vacuum Assisted Resin Transfer Molding.*" M.S. - Master of Science, Middle East Technical University (2018).

Ashcroft, I.A., Hughes D.J., and Shaw, S.J. "Adhesive Bonding of Fibre Reinforced Polymer Composite Materials." *Assembly Automation* 20, no. 2 (2006): 150–61.

Avendaño, R., Carbas, R.J.C., Marques, E.A.S., Da Silva, L.F.M. and Fernandes, A.A. "Effect of Temperature and Strain Rate on Single Lap Joints with Dissimilar Lightweight Adherends Bonded with an Acrylic Adhesive." *Composite Structures* 152 (2016): 34–44.

Ávila, A.F., and Bueno, P.D. "Stress Analysis on a Wavy-Lap Bonded Joint for Composites." *International Journal of Adhesion and Adhesives* 24, no. 5 (2004): 407–14.

Bagheri, R., and Pearson, R.A. "Role of Particle Cavitation in Rubber-Toughened Epoxies: 1. Microvoid Toughening." *Polymer* 37, no. 20 (1996): 4529–38.

Banea, M. D. and da Silva, L. F. "Adhesively Bonded Joints in Composite Materials: An Overview." Proceedings of the Institution of Mechanical Engineers, Part L: Journal of Materials: Design and Applications, January 1, 2009.

Banea, M.D., Rosioara, M., Carbas, R.J.C., and da Silva, L.F.M. "Multi-Material Adhesive Joints for Automotive Industry." *Composites Part B: Engineering* 151, no. May (2018): 71–77.

Bonpain, B. and Stommel, M. "Influence of Surface Roughness on the Shear Strength of Direct Injection Molded Plastic-Aluminum Hybrid-Parts." *International Journal of Adhesion and Adhesives* 82, no. January (2018): 290–98.

Boutar, Y., Naïmi, S., Mezlini, S. and Ali, M.B.S. "Effect of Surface Treatment on the Shear Strength of Aluminium Adhesive Single-Lap Joints for Automotive Applications." *International Journal of Adhesion and Adhesives* 67 (2016): 38–43.

Budhe, S., Banea, M. D., De Barros, S. and Da Silva, L.F.M., "An Updated Review of Adhesively Bonded Joints in Composite Materials." *International Journal of Adhesion and Adhesives* 72 (2017): 30–42.

Capelossi, V.R., Poelman, M., Recloux, I., Hernandez, R.P.B., De Melo, H.G. and Olivier, M.G. "Corrosion Protection of Clad 2024 Aluminum Alloy Anodized in Tartaric-Sulfuric Acid Bath and Protected with Hybrid Sol-Gel Coating." *Electrochimica Acta* 124 (2014): 69–79.

Carbas, R.J.C., Dantas, M.A., Marques, E.A.S. and da Silva, L.F.M. "Effect of the Adhesive Thickness on Butt Adhesive Joints under Torsional Loads." *Journal of Advanced Joining Processes* 3, no. December 2020 (2021): 100061.

Chowdhury, N.M., Chiu, W.K., Wang, J. and Chang, P. "Experimental and Finite Element Studies of Bolted, Bonded and Hybrid Step Lap Joints of Thick Carbon Fibre/Epoxy Panels Used in Aircraft Structures." *Composites Part B: Engineering* 100 (2016): 68–77.

Correia, S., Anes, V., and Reis, L. "Effect of Surface Treatment on Adhesively Bonded Aluminium-Aluminium Joints Regarding Aeronautical Structures." *Engineering Failure Analysis* 84, no. October 2017 (2018): 34–45.

Da Silva, L.F. and Lopes, M.J.C. "Joint Strength Optimization by the Mixed-Adhesive Technique." *International Journal of Adhesion and Adhesives* 29, no. 5 (2009): 509–14.

Da Silva, L.F., Rodrigues, T.N.S.S., Figueiredo, M.A.V., De Moura, M.F.S.F. and Chousal, J.A.G. "Effect of Adhesive Type and Thickness on the Lap Shear Strength." *Journal of Adhesion* 82, no. 11 (2006): 1091–1115.

Dhilipkumar, T. and Rajesh, M. "Effect of Using Multiwall Carbon Nanotube Reinforced Epoxy Adhesive in Enhancing Glass Fibre Reinforced Polymer Composite through Cocure Manufacturing Technique." *Polymer Composites*, no. April (2021): 1–15.

de Freitas, S.T., and Sinke, J. "Failure Analysis of Adhesively-Bonded Metal-Skin-to-Composite-Stiffener: Effect of Temperature and Cyclic Loading." *Composite Structures* 166 (2017): 27–37.

Gholami, R., Khoramishad, H. and da Silva, L.F. "Glass Fibre-Reinforced Polymer Nanocomposite Adhesive Joints Reinforced with Aligned Carbon Nanofillers." *Composite Structures* 253, no. May (2020): 112814.

Ghumatkar, A., Sekhar, R., and Budhe, S. "Experimental Study on Different Adherend Surface Roughness on the Adhesive Bond Strength." *Materials Today: Proceedings* 4, no. 8 (2017): 7801–9.

Giurgiutiu, V. "Introduction." *Structural Health Monitoring of Aerospace Composites*, 2016, 1–23.

Guo, L., Liu, J., Xia, H., Li, X., Zhang, X. and Yang, H. "Effects of Surface Treatment and Adhesive Thickness on the Shear Strength of Precision Bonded Joints." *Polymer Testing* 94 (2021): 107063.

Heidary, H., Karimi, N.Z. and Minak, G. "Investigation on Delamination and Flexural Properties in Drilling of Carbon Nanotube/Polymer Composites." *Composite Structures* 201, no. November 2017 (2018): 112–20.

Hu, X.F., Haris, A., Ridha, M., Tan, V.B.C. and Tay, T.E. "Progressive Failure of Bolted Single-Lap Joints of Woven Fibre-Reinforced Composites." *Composite Structures* 189, no. February (2018): 443–54.

Ishai, U.A. "Instantaneous effect of internal moisture conditions on strength of glass–fibre-reinforced plastics." In: Vinson, J.R., editor. *Advanced composite materials environmental effects*. Philadelphia: American Society for Testing and Materials; (1978). p. 267–76.

Jin, H., Mangun, C.L., Stradley, D.S., Moore, J.S., Sottos, N.R. and White, S.R.,"Self-Healing Thermoset Using Encapsulated Epoxy-Amine Healing Chemistry." *Polymer* 53, no. 2 (2012): 581–87.

Jojibabu, P., Zhang, Y.X. and Prusty, B.G. "A Review of Research Advances in Epoxy-Based Nanocomposites as Adhesive Materials." *International Journal of Adhesion and Adhesives* 96 (2020).

Kageyama M, Yoshida S. Development of XF-2 fighter composite structures, Development. 2000 Apr; 501:1364

Kanerva, M., and Saarela, O. "The Peel Ply Surface Treatment for Adhesive Bonding of Composites: A Review." *International Journal of Adhesion and Adhesives* 43 (2013): 60–69.

Karimi, N.Z., Heidary, H., Minak, G. and Ahmadi, M. "Effect of the Drilling Process on the Compression Behavior of Glass/Epoxy Laminates." *Composite Structures* 98 (2013): 59–68.

Kim, G.H., Choi, J.H. and Kweon, J.H. "Manufacture and Performance Evaluation of the Composite Hat-Stiffened Panel." *Composite Structures* 92, no. 9 (2010): 2276–84.

Kim, K.S., Yoo, J.S., Yi, Y.M. and Kim, C.G. "Failure Mode and Strength of Uni-Directional Composite Single Lap Bonded Joints with Different Bonding Methods." *Composite Structures* 72, no. 4 (2006): 477–85.

Kishore, A.N., and Siva Prasad, N. "An Experimental Study of Flat-Joggle-Flat Bonded Joints in Composite Laminates." *International Journal of Adhesion and Adhesives* 35 (2012): 55–58.

Kupski, J., and de Freitas, S.T. "Design of Adhesively Bonded Lap Joints with Laminated CFRP Adherends: Review, Challenges and New Opportunities for Aerospace Structures." *Composite Structures* 268, no. December 2020 (2021): 113923.

Kupski, J., De Freitas, S.T., Zarouchas, D., Camanho, P.P. and Benedictus, R. "Composite Layup e Ff Ect on the Failure Mechanism of Single Lap Bonded Joints." *Composite Structures* 217, no. December 2018 (2019): 14–26.

Li, X., Tan, Z., Wang, L., Zhang, J., Xiao, Z. and Luo, H. "Experimental Investigations of Bolted, Adhesively Bonded and Hybrid Bolted/Bonded Single-Lap Joints in Composite Laminates." *Materials Today Communications* 24, no. February (2020): 101244.

Lißner, M., Erice, B., Alabort, E., Thomson, D., Cui, H., Kaboglu, C., Blackman, B.R.K., Gude, M. and Petrinic, N. "Multi-Material Adhesively Bonded Structures: Characterization and Modelling of Their Rate-Dependent Performance." *Composites Part B: Engineering* 195, no. March (2020): 108077.

Lopes, J., Freitas, M., Stefaniak, D. and Camanho, P.P. "Inter-Laminar Shear Stress in Hybrid CFRP/Austenitic Steel" *Frattura ed Integrità Strutturale* 31, no. 67 (2015): 67–79.

Lucas, F M, and Adams, R. D. "Techniques to Reduce the Peel Stresses in Adhesive Joints with Composites" 27 (2007): 227–35.

Machado, J.J.M., Gamarra, P.M., Marques, E.A.S., and Lucas, F.M. "Improvement in Impact Strength of Composite Joints for the Automotive Industry." *Composites Part B* 138, no. November 2017 (2018): 243–55.

Mariam, M., Afendi, M., Majid, M.A., Ridzuan, M.J.M. and Gibson, A.G. "Tensile and Fatigue Properties of Single Lap Joints of Aluminium Alloy/Glass Fibre Reinforced Composites Fabricated with Different Joining Methods." *Composite Structures* 200, no. April (2018): 647–58.

Mohan, J., Ivanković, A., and Murphy, N. "Mixed-Mode Fracture Toughness of Co-Cured and Secondary Bonded Composite Joints." *Engineering Fracture Mechanics* 134 (2015): 148–67.

Moroni, F., Pirondi, A., Pernechele, C., Gaita, A., and Vescovi, L. "Comparison of Tensile Strength and Fracture Toughness under Mode I and II Loading of Co-Cured and Co-Bonded CFRP Joints." *Frattura Ed Integrita Strutturale* 13, no. 47 (2019): 294–302.

Mostafa, N.H., Ismarrubie, Z.N., Sapuan, S.M. and Sultan, M.T.H. "Fibre Prestressed Polymer- Matrix Composites: A Review." *Journal of Composite Materials* 51, no. 1 (2017): 39–66.

Naito, K., Onta, M., and Kogo, Y. "The Effect of Adhesive Thickness on Tensile and Shear Strength of Polyimide Adhesive." *International Journal of Adhesion and Adhesives* 36 (2012): 77–85.

Neto, J.A.B.P., Campilho, R.D. and Da Silva, L.F.M. "Parametric Study of Adhesive Joints with Composites." *International Journal of Adhesion and Adhesives* 37 (2012): 96–101.

Olajide, S.O., Kandare, E. and Khatibi, A.A. "Fatigue Life Uncertainty of Adhesively Bonded Composite Scarf Joints–an Airworthiness Perspective." *Journal of Adhesion* 93, no. 7 (June 7, 2017): 515–30.

Öz, Ö., and Özer, H. "An Experimental Investigation on the Failure Loads of the Mono and Bi-Adhesive Joints." *Journal of Adhesion Science and Technology* 31, no. 19–20 (2017): 2251–70.

Ozel, A., Yazici, B., Akpinar, S., Aydin, M.D. and Temiz, Ş. "A Study on the Strength of Adhesively Bonded Joints with Different Adherends." *Composites Part B: Engineering* 62 (2014): 167–74.

Panigrahi, S. K., and Pradhan, B. "Delamination Damage Analyses of Adhesively Bonded Lap Shear Joints in Laminated FRP Composites." *International Journal of Fracture* 148, no. 4 (2007): 373–85.

Park, Y.B., Song, M.G., Kim, J.J., Kweon, J.H. and Choi, J.H. "Strength of Carbon/Epoxy Composite Single-Lap Bonded Joints in Various Environmental Conditions." *Composite Structures* 92, no. 9 (August 2010): 2173–80.

Parker, B.M., "Some effects of moisture on adhesive-bonded CFRP–CFRP joints." *Composite Structures*, (1986):6:123–39.

Pascoe, J. A., Zavatta, N., Troiani, E., and Alderliesten, R. C. "The Effect of Bond-Line Thickness on Fatigue Crack Growth Rate in Adhesively Bonded Joints." *Engineering Fracture Mechanics* 229, no. February (2020): 106959.

Reddy, K.R., El-Zein, A., Airey, D.W., Alonso-Marroquin, F., Schubel, P. and Manalo, A. "Self-Healing Polymers: Synthesis Methods and Applications." *Nano-Structures and Nano-Objects* 23 (2020): 100500.

Rudawska, A. "Joint Strength of Hybrid Assemblies: Titanium Sheet-Composites and Aluminium Sheet-Composites -Experimental and Numerical Verification." *International Journal of Adhesion and Adhesives* 30, no. 7 (2010): 574–82.

Sancaktar, E., and Kumar, S. "Selective Use of Rubber Toughening to Optimize Lap-Joint Strength." *Journal of Adhesion Science and Technology* 14, no. 10 (2000): 1265–96.

Seong, M.S., Kim, T.H., Nguyen, K.H., Kweon, J.H. and Choi, J.H. "A Parametric Study on the Failure of Bonded Single-Lap Joints of Carbon Composite and Aluminum." *Composite Structures* 86, no. 1–3 (2008): 135–45.

Shanmugam, L., Naebe, M., Kim, J., Varley, R.J. and Yang, J. "Recovery of Mode I Self-Healing Interlaminar Fracture Toughness of Fibre Metal Laminate by Modified Double Cantilever Beam Test." *Composites Communications* 16, no. August (2019): 25–29.

Shin, K.C. and Lee, J.J. "Tensile Load-Bearing Capacity of Co-Cured Double Lap Joints." *Journal of Adhesion Science and Technology* 14, no. 12 (2000): 1539–56.

Shin, K.C., Lim, J.O. and Lee, J.J."The Manufacturing Process of Co-Cured Single and Double Lap Joints and Evaluation of the Load-Bearing Capacities of Co-Cured Joints." *Journal of Materials Processing Technology* 138, no. 1–3 (2003): 89–96.

Shokrian, M.D., Shelesh-Nezhad, K. and Najjar, R. "The Effects of Al Surface Treatment, Adhesive Thickness and Microcapsule Inclusion on the Shear Strength of Bonded Joints." *International Journal of Adhesion and Adhesives* 89, no. December 2018 (2019): 139–47.

Song, M.G., Kweon, J.H., Choi, J.H., Byun, J.H., Song, M.H., Shin, S.J. and Lee, T.J. "Effect of Manufacturing Methods on the Shear Strength of Composite Single-Lap Bonded Joints." *Composite Structures* 92, no. 9 (2010): 2194–2202.

Stevenson, B. "Comparison of the Mechanical Response of Co-Cured and Secondarily Bonded Composite Stiffened Panels". *The UNSW Canberra at ADFA Journal of Undergraduate Engineering Research* 8(1) (2015).

Taib, A.A., Boukhili, R., Achiou, S. and Boukehili, H. "Bonded Joints with Composite Adherends. Part II. Finite Element Analysis of Joggle Lap Joints." *International Journal of Adhesion and Adhesives* 26, no. 4 (2006): 237–48.

Temiz, Ş. "Application of Bi-Adhesive in Double-Strap Joints Subjected to Bending Moment." *Journal of Adhesion Science and Technology* 20, no. 14 (2006): 1547–60.

Walker, S.P. "Thermal effect on the compressive behavior of IM7/PET15 laminates." *Journal of Composite Material*, (2004):38:149–62

Watson, J. C., *Preliminary Design Development AV-B Fuselage Composite Structure in Fibrous Composites in Structural Design*, Boston, Springer, 1980.

Wisnom, M. R. "The Role of Delamination in Failure of Fibre-Reinforced Composites." *Philosophical Transactions of the Royal Society A: Mathematical, Physical and Engineering Sciences* 370, no. 1965 (2012): 1850–70.

Wu, C., Chen, C., He, L. and Yan, W. "Comparison on Damage Tolerance of Scarf and Stepped-Lap Bonded Composite Joints under Quasi-Static Loading" *Composites Part B: Engineering* 155, no. July (2018): 19–30.

Xiaoquan, C., Baig, Y., Renwei, H., Yujian, G. and Jikui, Z. "Study of Tensile Failure Mechanisms in Scarf Repaired CFRP Laminates." *International Journal of Adhesion and Adhesives* 41 (2013): 177–85.

Ye, Y., Zhu, W., Jiang, J., Xu, Q. and Ke, Y. "Computational Modelling of Postbuckling Behavior of Composite T-Stiffened Panels with Different Bonding Methods." *Composites Part B: Engineering* 166, no. November 2018 (2019): 247–56.

Zheng, G., Liu, C., Han, X. and Li, W. "Effect of Spew Fillet on Adhesively Bonded Single Lap Joints with CFRP and Aluminum-Alloy Immersed in Distilled Water." *International Journal of Adhesion and Adhesives* 99, no. March (2020): 102590.

Zielecki, W., Kubit, A., Kluz, R. and Trzepieciński, T. "Investigating the Influence of the Chamfer and Fillet on the High-Cyclic Fatigue Strength of Adhesive Joints of Steel Parts." *Journal of Adhesion Science and Technology* 31, no. 6 (2017): 627–44.

8 Impact Testing of Graphene, Epoxy and Carbon Fibre Reinforced Coir Composites and Automatic Prediction of Graphene Composition Using ANN

Tanay Kuclourya, Mohit Kumar Jain, Shubham Mudliar, Narendiranath Babu Thamba and Venkatesan S

Vellore Institute of Technology, Vellore, India

CONTENTS

DOI: 10.1201/9781003200994-8

87

8.1 INTRODUCTION

During the past several decades, natural materials which consist of cellulose-rich fibres have gained wide attention because of environmental problems like global climatic changes and pollution. Polymer composites derived from these fibres are always in huge demand. One such potential natural fibre is coir. From some studies, it has been found out that water absorption rate and void content in the coir-epoxy composite have a direct relation with the amount and the length of the fibres used (Das & Biswas 2016). In another study, it has been deduced that as the fibre weight percentage increases in the composite interface, the fibre-resin bond starts decreasing after reaching a particular point (Quy &Nguyen 2019).

Alkali such as NaOH is very commonly used to remove pectin. It attacks the pectin component and leaves no residue in the fibre after the treatment. It is also used to remove ether linkages by virtue of alkaline cleavages (Narendiranath Babu et al. 2019a). There is evidence supporting these theories. One such research result is enhancement in the mechanical properties of the poly butylene succinate resin when raw untreated fibres were replaced by alkali-treated coir fibres (Narendiranath Babu et al. 2019b). Such evidence is the increase in the flexural and impact strength of sisal-carbon hybridised composites due to the increase in the surface roughness topography by virtue of better adhesion of the fibres (Ahmad et al. 2019). However, having said so, chemical treatment if done excessively can lead to loss in the fibrous amount. Due to the damage in the composite after the testing, the carbon fibre breakage can be easily seen at several spots (Elzubair & Suarez, 2012). This can occur due to the excess of oxidant concentration, which affects the crystallinity of the end product (Narendiranath Babu et al. 2018ab, Khanam et al. 2010). To overcome this problem, a distilled water treatment can be done to remove the excess alkali. Some studies have shown that acidic treatment of fibres after removing excess alkali through distilled water can further increase the impact strength of the composite (Hallad et al. 2018). Presently in the world of materials and manufacturing engineering, the most reliable and viable materials for the fabrication of models are nano-particle–reinforced polymer composites. Carbon fibre is one such potential and very capable fibre used for reinforcing natural fibre composites. Many investigations have deduced that carbon fibre increases the strength of plain composites after reinforcement (Punyamurthy et al. 2013). Carbon fibre has an edge over glass fibre, bringing better mechanical properties to the composite (Anuar et al. 2007). However, carbon fibre reinforced polymer matrixes may have poor fracture toughness, but this can be improved by using organo-clay (Naveen et al. 2019). Today, carbon-fibre–reinforced epoxy resin composites have found their civil use in wide areas such as fuel cells, sports, bridge construction, etc. (Zhiyuan 2018).

Graphene (GO) is another very important reinforcing material due to its excellent thermal, mechanical and electrical properties. It has potential applications in the field of photonic and electronic devices (Borah et al. 2014). The presence of oxide groups in GO provides strong binding to the matrix and hence strength is enhanced (Pathak et al. 2016). For making graphene reinforced epoxy resin composites, graphene nano pellets (GnPs) are mixed into the epoxy by the sonication method. GO sheets can even increase the strength and interfacial rigidity of carbon-fibre–reinforced composites

(Zhang et al. 2012). One such study has shown that on CTAB treatment, there is a fine scattering of GO particles in the epoxy matrix (Saba et al. 2015, Ga 2019). A small concept of artificial intelligence (AI) has also been included in this research. Out of the various types of neural networks, this research seeks the advantage of only artificial neural networks (ANN). A relationship between input and output parameters has been developed after a keen analysis of all the minute details (Mohamed 2019, Ye et al. 2016). In this chapter, impact strength of composites made from coir, carbon fibre, graphene and epoxy has been determined. The mode of testing is Charpy test. Testing has been done on four different types of composites. All these composites vary in their graphene composition (1%, 2%, 4% and 6%). The fibres are oriented in uniaxial direction. The comparisons are done and the results are concluded. ANOVA as well as ANN analyses are done to check the validity and the authenticity of the data. Further, for finding the chemical and morphological details of the composites, FTIR and FESEM analyses are also done.

8.2 MATERIALS AND METHODS

8.2.1 MATERIALS

Coir fibres in bristle form are procured from fully ripped coconut husks by wet milling method and are used in this research work. For making the alkali solution fibres are treated with NaOH pellets. Graphene in the form of GO powder and carbon fibres are used as reinforcing materials. Chemicals which are used in this work are viscous epoxy resin (LY556), methanol, acetic acid; distilled water, cetyl trimethyl ammonium bromide (CTAB) and polyamine hardener (HY951) cured at room temperature. Poly vinyl alcohol (PVA) solution is also used as a mould-releasing agent. Galaxy Scientific Company, Vellore is the dealer for all the chemicals. All the fibres and reinforcing materials used in this research are shown in Figure 8.1.

8.2.2 METHODS

The following methodology has been adopted:

8.2.2.1 Chemical Treatment of Fibres

Fibres are washed with distilled water before performing chemical treatment to ensure the removal of suspended dust particles. Then fibres are treated with 5% (by weight) NaOH solution for two hours to improve surface morphology. After NaOH treatment fibres are dipped into acetic acid until the pH reaches 7, which is determined using litmus paper. After acidic treatment fibres are again washed with distilled water. Then fibres are left for drying at room temperature. Fibres are then ready to be used for making composites.

8.2.2.2 Sonication and Incubation

Four different beakers were taken to prepare four different compositions of composites with varying graphene percentage by 1%, 2%, 4% and 6% by weight. For making compositions, 100 ml methanol is mixed with 2 grams of graphene and 1

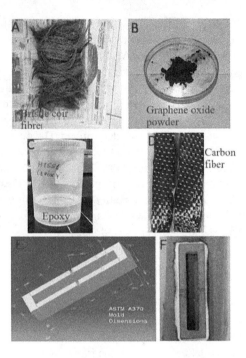

FIGURE 8.1   (a) Bristle coir fibre, (b) Graphene, (c) Epoxy, (d) Carbon fibre, (e & f) Mould (A-370).

gram of C-Tab (depending on the percentage by weight composition of graphene). It takes around 2 hours for complete sonication of mixture. The sonicated mixture is combined with 200 ml of epoxy and for better mixing it is put inside an incubator for continuous stirring at 50°C until the methanol evaporates completely. The treated fibres are then dipped inside the mixture obtained after incubation. To act as a mould releasing agent, 5% aqueous solution of PVA (poly vinyl alcohol) is poured in the mould cavity. Now hardener is added to the treated fibre mixture and stirred for about 2–3 minutes for even mixing. The fibres, along with the epoxy and graphene mixture, are arranged in uniaxial orientation in the mould using hand lay-up technique. These specimens are ready for the removal process after drying for 20–25 days, depending on the weather conditions. The ideal temperature and pressure is atmospheric pressure at 31°C. Removal of the specimen is carried out with utmost care in order to prevent damaging the specimens. The dried specimens are then taken for Charpy impact testing.

8.3   RESULTS AND DISCUSSION

In this research work, Charpy testing has been followed as a standard to measure the impact strength of the composites. The results can be analysed on the basis of different parameters such as graphene composition and chemical properties. Table 8.1 shows the impact results obtained.

TABLE 8.1

Results from Charpy Impact Test

Sr. No.	Specimen label (% wt. by graphene)	Charpy Impact Strength (J)						
		S1	S2	S3	S4	S5	Mean	Standard Deviation
1	1%	117.8	112.2	108.7	111.6	121.7	**114.4**	4.69
2	2%	150.1	148.6	141.3	145.5	154.5	**148.0**	4.43
3	4%	133.9	123.9	122.6	127.3	134.8	**128.5**	5.03
4	6%	139.8	133.5	132.2	140.4	143.1	**137.8**	4.21

8.3.1 COMPARISON ON THE BASIS OF GRAPHENE COMPOSITION

From the results as shown in Table 8.1, the average impact strength values of 1%, 2%, 4% and 6% graphene composites were found to be 114.4, 148, 128.5 and 137.8 J, respectively. It is evident that impact strength enhances on increasing graphene composition in the composites. However, 2% graphene shows exceptionally high impact strength. Five samples have been tested for each percentage composition in order to get reliable results. S1, S2, S3, S4 and S5 are the terms used for distinguishing the five samples.

The presence of more graphene leads to better bonding in the matrix, which ultimately results in results in better reinforcement. Previous studies have also proved that graphene increases the strength and stiffness of the composites. The results indicate that 2% is the ideal percentage composition which can withhold maximum impact strength. Because of the synergistic effect, 2% of graphene is optimum with carbon fibre and coir. This effect takes place through the interaction of carbon fibre and graphene. Carbon fibre shows phenomenal characteristics because of its interior graphite crystallites which contain graphene in an interconnected manner, as well as in a turbostratic manner. In this structure, the size and alignment play a major role in enhancing the performance. Graphite scaling of the size of crystallite eliminates the grain boundaries to the utmost extent in the lateral direction, which causes the breakthrough and hence optimum results were obtained. It should be noted that synergistic effect and strength are interconnected. Both factors influence one each other.

8.3.2 ANALYSIS OF VARIANCE (ANOVA)

Two-way analysis of variance (ANOVA) has been carried out for the statistical analysis of the data such as the flexural strength, ultimate tensile strength and impact strength of the composites. These data are classified on the basis of two factors: percentage composition of graphene and the type of mould used. The data for flexural and tensile strength has been derived from a recent work of the authors and is shown in Table 8.2. A two-way ANOVA table has been made and the test results are shown in Table 8.3, after the formulation of data. ANOVA analysis as shown in Table 8.3 decomposes the variance of the strength of the composites into two factors. All

TABLE 8.2

Data Values

Percentage Composition of graphene (Factor-1)	Type of mould (Factor-2)		
	Flexural (MPa)	Tensile (MPa)	Impact (J)
1%	110.89935	29	114.4
2%	405.08408	50	148
4%	281.78925	38	128.5
6%	373.73444	43	137.8

TABLE 8.3

ANOVA Test for Testing Effect of Factors on Strength of the Composites

Source of Variation	Sum of squares (SS)	Degrees of freedom (df)	Mean Square (MS)	F (Experimental)	F (Critical)	p
Percentage Composition of graphene	47838.42096	3	15946	3.595	3.49	0.046275
Type of mould	262047.88952	2	131204	29.543	3.89	0.000023
Interaction	58586.94	6	9765	2.202	3.00	0.11523
Within	53212.6852	12	4435			
Total	421685.9357	23				

the results are validated at 95% confidence level ($\alpha = 0.05$). The p values for factor-1 and factor-2 are 0.046275 and 0.000023, respectively, which is less than 0.05. Hence according to the test, the hypotheses neglecting the significant change of these factors on the data are rejected. On the other hand, p value for the interaction of these two factors is 0.11523, which is greater than 0.05. Hence the third hypothesis neglecting the significant effect of this factor on the data is accepted. Hence, statistically, the percentage composition of graphene has significant effect on the strength of the composites. Similarly, the type of the mould used also significantly affects the strength of the composites. However, the interaction of these two factors has no significant effect on the strength of the composites [23–24].

8.3.3 ARTIFICIAL NEURAL NETWORKS (ANN)

An artificial neural network (ANN) has been trained by the impact test data. A pattern has been established between the input and the output parameters. After the comparison of the targeted and the actual output, accuracy of the ANN has been found along with some more interesting inferences. For this, impact strength has been taken as an input parameter and percentage composition of graphene used in the composites is taken as the output parameter. Since the MATLAB software demanded the output data in a binary format, some distinct binary values have been allocated to each of the four weight percentages of graphene (1% = 0001, 2% = 0010, 4% = 0100 and 6% = 1000). For thorough analysis, several plots such as the confusion matrix, performance diagram, receiver operating characteristic (ROC) plot, training performance diagram and error histogram have been plotted as shown in Figure 8.2. For training purposes, neural pattern recognition tool (nprtool) is used. The nprtool is present in the neural network toolbox in MATLAB. For constructing the pattern, 10 hidden neurons have been used. These neurons are provided by the Scaled Conjugate Gradient back-propagation (trainscg) tool. Of the total impact test data

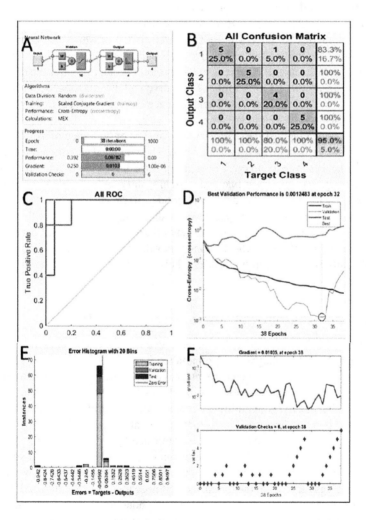

FIGURE 8.2 (a) Data obtained from the Artificial Neural Network (ANN) trained on the Impact testing data; (b) Summary of the confusion matrix; (c) Receiver Operating Characteristic (ROC) diagram; (d) Performance plot; (e) Error Histogram; (f) Training state performance diagram. ANN Analyses.

provided to both the networks, 70% has been utilised by the network to train itself, 15% has been utilised for validation purposes and the remaining 15% has been used to check the accuracy of the networks. The various plots mentioned above have their own significance. The confusion matrix shows the accuracy of the network by classifying the data as correct, false positive or false negative. The main diagonal of the confusion matrix represents the correct output whereas the values are the false positive and the false negative values. The overall accuracy of the network is the cumulative probability of correct output predicted for each class of outcome. In this research, the accuracy of the network is 95%. An epoch is one complete cycle of iteration performed by the network while testing the data. With the passing of each epoch, there is a drop in the errors. This measurement is done by the performance plot. Once the 1000th epoch is reached or the lowest error has occurred, the training process automatically stops. The ROC diagram is a plot which shows the relationship between the true and false positive rates and hence checks the performance of the network. The nearer the curve is to the top-left border, the better the performance of the network. Similarly, proximity of the curve with the diagonal denotes poor performance of the trained network. The difference between the actual and the targeted output by the network is known as error.

The errors can be positive, zero or negative. The entire range of errors is divided into 20 small segments which are called bins. The graphical classification of data into these 20 bins is shown in an error histogram. If the tallest bin in the error histogram is very close to the zero error line and is significantly taller than the rest of the bins, then the network has high accuracy and good performance. Finally, the last plot of this analysis is the training state performance diagram. The first part of the plot shows the gradient of the loss function of the network, whereas the second part shows the scatter plot of the validation checks. Once six validation checks are failed, MATLAB automatically stops the training process. This inbuilt feature prevents the overtraining of the network.

8.3.4 FOURIER TRANSFORM INFRARED SPECTROSCOPY (FTIR)

FTIR analysis is an analytical technique to determine surface chemical properties of specimens. FTIR analysis was conducted by choosing mid IR range of 400–4000 cm^{-1}. FTIR helps to detect the presence of various bonds and material present inside the specimen differentiated by the various wt% of graphene. Figure 8.3 shows the FTIR analysis of all the specimens. Alkali treatment lead to the reduction in hydrogen bonding because it leads to the removal of hydroxyl groups which reacts with sodium hydroxide NaOH, which is confirmed by a decrease in intensity of the peak between 3300 and 3320 cm^{-1} (Abraham et al. 2013). A wide and strong point value of 3340 cm^{-1} is shown by 6% impact results, which represent stretching vibrations of O-H resulting from the lignin and cellulose texture of the fibre. All four cases show no presence of 1728cm^{-1} as it shows C=O bond axial vibration which is missing after chemical treatment of fibres. This proves that coir fibres are treated with alkali. A band showing 1240cm^{-1} confirms the presence of C-O vibrations from phenols, esters and ethers, which happens because of waxes in the epidermal tissue; no such band present proves that there isn't any removal of wax in the treated

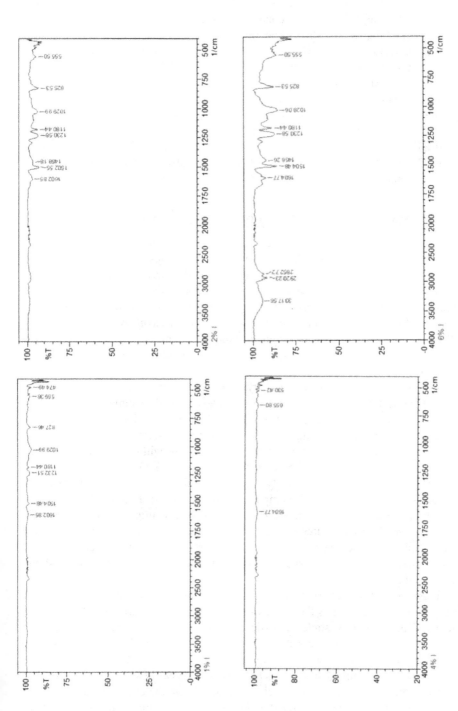

FIGURE 8.3 Fourier transform infrared spectroscopy analysis of different weight percentages of graphene (1%, 2%, 4%, 6%).

fibres (Sari et al. 2015). Studies have shown that bands at 1459 cm^{-1} represent the bending vibrations of group CH3 and CH bonds. Peaks at 2852cm^{-1} and 2893 cm^{-1} are the stretching vibrations of bond CH (Song et al. 2015). In another study, it was observed that bands which show frequency between 3200 and 3600 cm^{-1} show stretching vibrations of O-H bond. Bands at 1461cm^{-1} show ammonium group–stretching vibrations. 1446 cm^{-1} shows the typical CTAB absorption bands (Shyam Suraj et al. 2020). From various analyses in previous studies, it is inferred that aromatic carbon double bond is proved by the band presence at 1250 and 1600 cm^{-1}. Presence of C-O stretching bonds is confirmed by the presence of bands between 1050–1150 cm^{-1}. Also, 2850–3000 cm^{-1} show the presence of stretching bonds of C-H alkane as there are peaks defining it. Band frequencies 801cm^{-1} and 821 cm^{-1} show C-H bond. On the other hand, 1234 cm^{-1} and 1501 cm^{-1} proves the presence of C-C bond and aromatic nitro compounds (El Khadem 1975).

8.3.5 Scanning Electron Microscopy (SEM) Analysis

Scanning electron microscopy (SEM) is a method used to analyse the morphological details of the nanostructures. SEM images provide various details which cannot be analysed with the naked eye. Minute details such as the size and shape of the agglomerated fibre sheets can also be visualised in the composites. The structures can be seen in the form of rods, layers, flakes, etc. Field emission scanning electron microscopy (FESEM) is one type of SEM method which is carried out in this research work. The specimens, sized around 10mm × 10mm × 10mm, are extracted from every composite before and after the testing. These cut out specimens are placed in holders called STEM holders.

In order to increase the conductivity of the specimens, coating of a gold-palladium layer is done on the specimen surface. These coated specimens are then taken for FESEM analysis. In this research work, FESEM analyses have been done in two phases – before impact testing and after impact testing. Before testing, the focus is on detecting the presence of different entities present within the composites and how the composite reacts to Charpy test (impact testing). On the other hand, after testing, the main focus is on analysing fibre breakage and void generation. The presence of coir fibres in the composites can be seen in Figure 8.4a.The figure clearly shows that the fibres are arranged in uniaxial direction and are present in the form of long strands. Figures 8.4b and 8.4c show the rod-like structures which are carbon fibres. These figures also depict the carbon fibre arrangement inside the composites. The flake-like structures which are seen over the carbon fibres shows the presence of GO.

Figure 8.4d shows a very important image of a sonicated mixture of graphene and epoxy. The size as well as the shape of the sheets can be clearly depicted from the image. Other than this, some small-sized sheets are also visible in the image. The layered types of structure are the agglomerated sheets. FESEM analysis is again done on the composites once the impact testing is over. Due to the damage in the composite after testing, carbon fibre breakage can be easily seen at several spots in Figure 8.5a. The fibres are completely disoriented from the usual uniaxial direction. Voids and cracks are also generated in the GO layer by virtue of the impact damage

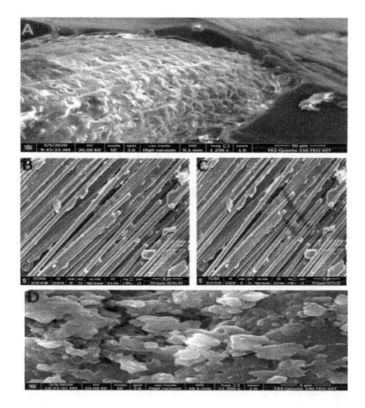

FIGURE 8.4 SEM images before testing (a) coir fibre, (b & c) carbon fibre rod-like structure, (d) graphene flakes.

within the composite. These cracks and voids can be seen in Figures 8.5b and 8.5c. Further, Figure 8.5d also shows coir fibre breakage. The coir fibre surface can be seen covered by tiny bits of GO flakes, which are wavered in many directions (Kuclourya Tanay et al. 2020, Michaeli & Londschien 1991, Hernandez et al. 2017).

8.4 CONCLUSION

The impact strength of the coir composites reinforced with graphene, epoxy and carbon fibre have been investigated for different weight percentages of graphene (1%, 2%, 4% and 6%). It was observed that the impact strength of the composites was in the decreasing order of 2% > 6% > 4% > 1%. Apart from 2% graphene composite, which gives the optimum result, the general trend observed is on increasing graphene, as there is an increase in impact strength. There is a percentage increase of over 20% in impact strength when graphene composition is increased from 1% to 6%. Impact strength is clearly governed by the amount of graphene present in the composite. The results conclude that 1% composition is very rigid and brittle in nature, whereas 2% can bear maximum stress while being extended. Statistical

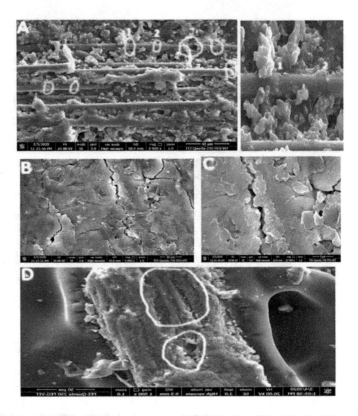

FIGURE 8.5 Cracks in SEM images after testing (a) carbon Fibre Rods, (b & c) graphene flake structure, (d) coir Fibre.

analysis (ANOVA) concluded that factors such as percentage composition of graphene and type of mould brought significant change to the results individually but the interaction of these two factors had no significant effect on the strength of the composites. For checking the authentication of the impact test data, neural network models such as artificial neural network (ANN) were trained using the data and the accuracy of these models was analysed. A pattern was constructed between the input parameter (impact strength) and output parameter (percentage composition of graphene). The targeted as well as the actual output values were compared and then the accuracy result of the network was obtained. It was observed that the ANN network exhibited 95% accuracy. The good results in accuracy prove that the impact test data is very promising. Finally, various functional groups were detected in the composites through FTIR analysis. The FTIR graph displayed different wave numbers, which indicated the presence of different functional groups and chemical reactions which took place within the composite. Organic solvents were also detected, which further proved that different chemical structures lead to different chemical reactions within the specimen. FESEM analysis was also carried out to show the presence of all the constituents of the composites such as coir fibres, epoxy, graphene and carbon fibres. Fibre breakage, voids and crack generation were also shown by conducting FESEM on specimens which underwent the impact test. Different sectors such as automobiles, defense, electronics and packaging can significantly rely on coir-graphene-carbon-epoxy composites because of their light weight and high strength. However, like any other research, this research work too leaves behind some stones unturned, such as

- Performing wear test on composites for finding the resistance towards wear.
- Performing ballistic test on composites to find toughness.
- Finding electrical and thermal characteristics of the composites.
- Performing XRD analysis on composites for determining the degree of crystallinity of the composite.

REFERENCES

Abraham, Eldho, B. Deepa, L.A. Pothen, J. Cintil, S. Thomas, M.J. John, R. Anandjiwala, and S.S. Narine. "Environmental Friendly Method for the Extraction of Coir Fibre and Isolation of Nanofibre." Carbohydrate Polymers 92, no. 2: 1477–83, 2013

Ahmad, Mohammad Ayaz, Gülşen Güven Güven, and Nursabah Sarıkavaklı. "Some Features of Doping of Nano–Graphite in Natural Coir Fibre Epoxy Composites." European Journal of Science and Technology, 491–98, 2019

Anuar, H., S. H. Ahmad, R. Rasid, A. Ahmad, and W. N. Wan Busu. "Mechanical Properties and Dynamic Mechanical Analysis of Thermoplastic-Natural-Rubber-Reinforced Short Carbon Fibre and Kenaf Fibre Hybrid Composites." Journal of Applied Polymer Science 107, no. 6: 4043–52, 2007

Borah, Munu, Minakashi Dahiya, Shaveta Sharma, R. B. Mathur, and Sanjay R. Dhakate. "Few Layer Graphene Derived from Wet Ball Milling of Expanded Graphite and Few Layer Graphene Based Polymer Composite." Materials Focus 3, no. 4: 300–309, 2014

Das, Geetanjali, and Sandhayarani Biswas. "Effect of Fibre Parameters on Physical, Mechanical and Water Absorption Behaviour of Coir Fibre–Epoxy Composites." Journal of Reinforced Plastics and Composites 35, no. 8, 2016

El Khadem, Hassen S. "Spectrometric Identification of Organic Compounds. Third Edition (Silverstein, Robert M.; Bassler, G. Clayton; Morrill, Terrence C.)." Journal of Chemical Education 52, no. 6, 1975

Elzubair, Amal, and João Carlos Miguez Suarez, "Mechanical Behavior of Recycled Polyethylene/Piassava Fibre Composites.", Materials Science and Engineering: A 557:29–35, 2012

Ga, Korablev. "Enriched Mechanical Properties of Epoxy/Coir Fibre Composites with Graphene Oxide." Research & Development in Material Science 10, no. 5, 2019

Hallad, Shankar A., N.R. Banapurmath, Vishal Patil, Vivek S Ajarekar, Arun Patil, Malatesh T. Godi, and Ashok S. Shettar. "Graphene Reinforced Natural Fibre Nanocomposites for Structural Applications." IOP Conference Series: Materials Science and Engineering 376: 012072, 2018

Hernandez, Dany Arnoldo, Carlos Alberto Soufen, and Marcelo Ornaghi Orlandi. "Carbon Fibre Reinforced Polymer and Epoxy Adhesive Tensile Test Failure Analysis Using Scanning Electron Microscopy." Materials Research 20, no. 4: 951–61, 2017

Khanam, P. Noorunnisa, H. P. S. Abdul Khalil, M. Jawaid, G. Ramachandra Reddy, C. Surya Narayana, and S. Venkata Naidu. "Sisal/Carbon Fibre Reinforced Hybrid Composites: Tensile, Flexural and Chemical Resistance Properties." Journal of Polymers and the Environment 18, no. 4: 727–33, 2010

Michaeli, W., and M. Londschien. "Plasma Treatment To Improve Fibre/Matrix-Adhesion." *Interfacial Phenomena in Composite Materials '91*, 30–33, 1991

Mohamed, Zahraa E. "Using the Artificial Neural Networks for Prediction and Validating Solar Radiation." Journal of the Egyptian Mathematical Society 27, no. 1, 2019

Narendiranath Babu, Thamba, Singh Robin, Dogra Sanchit, "Wear Characteristics of Epoxy Resin Based Composites Reinforced With Aloe Fibres in Combination with Al_2O_3/SiC". Materials Today: Proceedings 5:12649–12656, 2018a

Narendiranath Babu, Thamba, Dogra Sanchit, Chauhan Milind, Singh Robin, Shah Jaineel, Irshad Alam M D, Kumar Rai Hemant, Kothari Kshitij, Ranjan Gupta Shubhankar," A Review On Aloe/Kapok/Palmyra/Corn Fibre/Vetivernatural Fibres For Biomedical Applications". Materials Today: Proceedings 5 13535–13546, 2018b

Narendiranath Babu, Thamba, Chauhan Milind, Alam Md Irshad, Shah Jaineel." Improving Wear Properties Of Musti Kusa Grass Fibre Reinforced With Aluminium Oxide And Silicon Carbide". Materials Today: Proceedings 5: 13003–13009, 2018c

Narendiranath Babu, Thamba, Kuclourya Tanay, Jain Mohit Kumar, Ramalinga Viswanathan Mangalaraja, "Flexural properties of areca nut, sunn hemp and e-glass fibres reinforced with epoxy composites" Materials Research Express 6(10), 2019a

Narendiranath Babu, Thamba, Kuclourya Tanay, Jain Mohit Kumar, Ramalinga Viswanathan Mangalaraja, "Sunn Hemp Natural Fibres and E-Glass Fibres Reinforced With Epoxy Composites." International Journal of Engineering and Advanced Technology (IJEAT), 8 6, 2019b

Naveen, J., M. Jawaid, E. S. Zainudin, Mohamed T. H. Sultan, and R. Yahaya. "Improved Interlaminar Shear Behaviour of a New Hybrid Kevlar/Cocos Nucifera Sheath Composites with Graphene Nanoplatelets Modified Epoxy Matrix." Fibres and Polymers 20, no. 8: 1749–53, 2019

Pathak, Abhishek K., Munu Borah, Ashish Gupta, T. Yokozeki, and Sanjay R. Dhakate. "Improved Mechanical Properties of Carbon Fibre/Graphene Oxide-Epoxy Hybrid Composites." Composites Science and Technology 135: 28–38, 2016

Punyamurthy, Ramadevi, Dhanalakshmi Sampathkumar, Basavaraju Bennehalli, and Chikkol V Srinivasa. "Influence of Esterification on the Water Absorption Property of Single Abaca Fibre." Chemical Science Transactions 2, no. 2: 413–22, 2013

Saba, Naheed, Mohammad Jawaid, Othman Y Alothman, Mt Paridah, and Azman Hassan. "Recent Advances in Epoxy Resin, Natural Fibre-Reinforced Epoxy Composites and Their Applications." Journal of Reinforced Plastics and Composites 35, no. 6: 447–70, 2015

Sari P.S., Petr Spatenka, Zdenka Jenikova, Yves Grohens, and Sabu Thomas. "New Type of Thermoplastic Bio Composite: Nature of the Interface on the Ultimate Properties and Water Absorption." RSC Advances 5, no. 118: 97536–46, 2015

Song, Jianbin, Quanping Yuan, Xueshen Liu, Dong Wang, Feng Fu, and Wenbin Yang. "Combination of Nitrogen Plasma Modification and Waterborne Polyurethane Treatment of Carbon Fibre Paper Used for Electric Heating of Wood Floors." BioResources 10, no. 3, 2015

Suraj, Shyam, Kaul Shivam, Kalsara Nirav and Thamba Narendiranath Babu," "Mechanical behaviour and microscopic analysis of epoxy and E-glass reinforced banyan fibre composites with the application of artificial neural network and deep neural network for the automatic prediction of orientation", Journal of Composite Materials 0(0) 1–22, 2020-10-15

Tanay, Kuclourya, Jain Mohit Kumar, Mudliar Shubham and Thamba Narendiranath Babu, "Statistical analysis and investigation of tensile test data of coir composites reinforced with graphene, epoxy and carbon fibre", Proc IMechE Part L:J Materials: Design and Applications 0(0) 1–12, 2020

Van Quy, Ho, and Song Thanh Thao Nguyen. "Experimental Analysis of Coir Fibre Sheet Reinforced Epoxy Resin Composite." IOP Conference Series: Materials Science and Engineering 642: 012007, 2019

Ye, Chengxi, Chen Zhao, Yezhou Yang, Cornelia Fermüller, and Yiannis Aloimonos. "LightNet." *Proceedings of the 2016 ACM on Multimedia Conference - MM '16*, 2016.

Zhang, Xiaoqing, Xinyu Fan, Chun Yan, Hongzhou Li, Yingdan Zhu, Xiaotuo Li, and Liping Yu. "Interfacial Microstructure and Properties of Carbon Fibre Composites Modified with Graphene Oxide." ACS Applied Materials & Interfaces 4, no. 3: 1543–52, 2012

Zhiyuan, Yu. "Carbon Fibre Reinforced Epoxy Resin Matrix Composites." Materials Science: Advanced Composite Materials 1, no. 1, 2018

9 Failure of Composite Structure

Modelling of Cracks in the Aerospace Composite Structure

Muthukumaran Gunasegeran and Edwin Sudhagar P

Vellore Institute of Technology, Vellore, India

Ananda Babu A

Sharda University, Greater Noida, India

CONTENTS

DOI: 10.1201/9781003200994-9

9.1   INTRODUCTION

Composite materials use has increased day by day in many areas such as aircraft, aviation and automobiles. Even with high manufacturing cost composite materials are preferred to conventional materials, aluminium and titanium alloys as their strength-to-weight ratio is higher and they have a high fatigue characteristic specific strength and stiffness, as well as corrosion and temperature resistance, radar absorption and suppression (Gopal, 2016; Hiken, 2018; Kassapoglou, 2013; Peters, 1998; Soutis, 2005a, 2005b). Given the rise in the use of composite materials, the material behaviour and modes of failure need to be addressed through the studies and experiments carried out.

In this chapter, we discuss the different kinds of failure in composite structures and the modelling of delamination in plate structure.

9.2   FAILURE ANALYSIS OF COMPOSITE MATERIALS

Composite materials failure study is similar to other materials, therefore the same method can be adopted for composite materials. However, failure analysis is more difficult than a comparison of composite materials to metallic structures. The complexity of a combined material failure analysis was caused by material anisotropy, the application of adhesive bonds and the ability to crack in different modes of failure (Zimmermann & Wang, 2020).

9.2.1   TENSILE FAILURE

In this research by Mostafa et al. (2016), composites from woven E-glass/polyester with various presetting levels in equipment were prepared. Tests of fatigue were carried out to establish the influence of cyclic loading from the fabric pretension on the life of the pre-pressed composite. The fatigue life was increased to ~ 43% with the composite system shown in this work.

9.2.2   COMPRESSION FAILURE

In (Daniel et al., 2002; Gdoutos et al. 2003), sandwich columns and compression beams were widely examined for their facial wrinkling failures. The beams are made of the unidirectional ply of carbon/epoxide and aluminium honeycomb and cell-closed foam PVC panels. Where the thickness of the beam and the distance are short, a wrinkling failure was found to be usual.

Opelt et al. (2018) investigated the inadequately studiedfractographical aspects of compression defects, primarily due to losses on the fracture surfaces produced by the compressive load. The suggested classification was based on a literature analysis and microscopic experiments. On fibre-reinforced polymer composites, the wedge-splitting failure mode was first described.

9.2.3   SHEAR FAILURES

Shear failures of composites are very similar to that of homogeneous materials. Based on the sequence of stacking applied, the spread of the failure could be the subsequent fibre of an off-axis layer enhancing the adjacent layers fail due to the shear (Opelt et al., 2018).

9.2.4 BENDING OF LAMINATES

The three-point bending test was applied on the surface, resulting in tensile and compressive fracture. the following characteristics were presented (Shikhmanter et al., 1989).

9.2.5 IMPACT

Damage due to impact reduces the carrying capacity and residual compression, stiffness and stability of sandwich structures (Zimmermann & Wang, 2020).

9.2.6 FATIGUE

Belingardi et al. (2007) performed an experiment to assess the fatigue behaviour of wave sandwich beam. Two separate failure mechanisms were found: undamaged specimen failure caused by compressed faces collapsing, and the honeycomb cell walls breaking on the tip of the deep part of the specimen failure (Zimmermann & Wang, 2020).

9.3 FAILURE OF SANDWICH STRUCTURE

Daniel et al. (2002), and Gdoutos et al. (2003) found five modes for disruptions, including compression failure, wrinkling of face sheet and indentation failure that characterize the composite sandwich beams. (Zimmermann & Wang, 2020)

9.3.1 COMPRESSION OF SANDWICH STRUCTURES

Where the axially compressive force is present in a sandwich panel, the potential failure modes face compressive failure, facial wrinkle, global buckling and core shear instability (Gdoutos et al., 2003). Compressive failure occurs where the axial compression force is equal to the compressive pressure of the face material. In comparison, face wrinkling is defined if the tension in the face exceeds a certain critical value (Zimmermann & Wang, 2020).

9.3.2 BENDING OF SANDWICH STRUCTURES

Daniel et al. (2002) and Gdoutos et al. (2003) noted that when exposed to a pure bending force, the sandwich beams will fail in two ways, likewise described as four-point bending: face shrinkage and compressive failure. Compressive failures occur where the face tension is equal to the compressive strength of the materials. Compressive failures in the direction of thickness are experienced by structures with highly stiffened cores. They describe four modes of failure: shear failure of core, mixed shear core failure of core and compression, facing strain and compressive failure. Facing wrinkling failure will first be experienced with low shear forces; conversely with strong shear forces, core shear failure will first occur. Following the main shear failure are compressive and wrinkling faults (Zimmermann & Wang, 2020).

9.4 MODELLING OF DELAMINATION CRACK IN THE COMPOSITE PLATE STRUCTURE USING FEM

Delaminated composite components by distinct delamination positions are examined on delamination edge and embedded mid-plan delamination in the development of delamination modelling in composite. The entire delamination is divided into (d_n) axial area number (number) and horizontally strengthening laminas are uniformly retained in the x direction. A laminated plate's thickness (t_z) is considered to be quite tiny compared to the length of plates L and W of delaminated composite constructions. By examining delaminated portions 1 and 2, as clearly illustrated in Figure 9.1, the displacement function of delaminated portions 3 and 4 was derivable of continuity. For the delaminated segments top (1) and bottom (2), relative corresponding to the mid-plane, the local co-ordinates are $\{x_{11}, y_{11}, z_{11}\}$ and $\{x_{22}, y_{22}, z_{22}\}$, respectively.

The mid-plane deformation of the intact plate of $\{u_1^{mp}, u_2^{mp}, u_3^{mp}\}$ for the portion 3 and 4 is not equivalent, agreeing to equivalent single-layer theory assumptions, to the centre-plane deformations $\{u_1^{mpl}, u_2^{mpl}, u_3^{mpl}\}$ and $\{u_1^{mp2}, u_2^{mp2}, u_3^{mp2}\}$ of de-bonded plate region 1 or 2, both in their order (Arun Kumar & Mallikarjuna Reddy, 2020).

Given the un-delaminated region thickness (h_i), the top and bottom delaminated portion of the plate are h_{11} and h_{22} respectively. Where: mp-mid-plane, 1-top, 2-bottom.

Displacement due to mid-plane delamination might be explored in order to maintain the continuity between part 3 (intact) and region 1 (delaminated), as indicated in the following Equation 9.1:

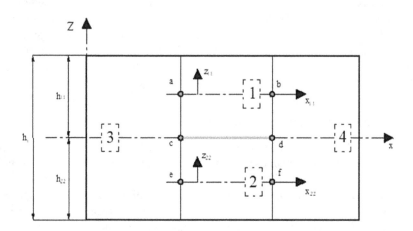

FIGURE 9.1 Midplane delaminated region of a plate – Side view.

$$\left\{ \begin{array}{l} u_1^{mpl}\left(x_{11},y_{11},t\right)=u_1^{mp}\left(x,y,t\right)-z_{11}\dfrac{\partial u_3^{mp}}{\partial x} \\[3mm] u_2^{mpl}\left(x_{11},y_{11},t\right)=u_2^{mp}\left(x,y,t\right)-z_{11}\dfrac{\partial u_3^{mp}}{\partial y} \end{array} \right\} \tag{9.1}$$

The mid-plane displacement due to delamination can be examined to maintain the continuity between part 3 (intact) and area 2 (delaminated) as illustrated in the following Equation 9.2:

$$\left\{ \begin{array}{l} u_1^{mp2}\left(x_{22},y_{22},t\right)=u_1^{mp}\left(x,y,t\right)-z_{22}\dfrac{\partial u_3^{mp}}{\partial x} \\[3mm] u_2^{mp2}\left(x_{22},y_{22},t\right)=u_2^{mp}\left(x,y,t\right)-z_{22}\dfrac{\partial u_3^{mp}}{\partial y} \end{array} \right\} \tag{9.2}$$

In their sequence the space between the centre of the plate of areas 3 and 1, 3 and 2 are $\left\{ z_{11} = \dfrac{h-h_{22}}{2} \right\}$ and $\left\{ z_{22} = \dfrac{h-h_{11}}{2} \right\}$

Equation 9.3 gives delaminated plate part 1 displacement:

$$\left\{ \begin{array}{l} u_1^{11}\left(x_{11},y_{11},z_{11},t\right)=u_1^{mpl}\left(x_{11},y_{11},t\right)-z_{11}\left(x\right)\dfrac{\partial u_3^{mp}}{\partial x} \\[3mm] u_2^{11}\left(x_{11},y_{11},z_{11},t\right)=u_2^{mpl}\left(x_{11},y_{11},t\right)-z_{11}\left(x\right)\dfrac{\partial u_3^{mp}}{\partial x} \\[3mm] \quad\quad u_3^{11}\left(x_{11},y_{11},z_{11},t\right)=u_3^{mp}\left(x,y,t\right) \end{array} \right\} \tag{9.3}$$

Equation 9.4 provides delaminated plate portion 2 displacement:

$$\left\{ \begin{array}{l} u_1^{22}\left(x_{22},y_{22},z_{22},t\right)=u_1^{mp2}\left(x_{22},y_{22},t\right)-z_{22}\left(x\right)\dfrac{\partial u_3^{mp}}{\partial x} \\[3mm] u_2^{22}\left(x_{22},y_{22},z_{22},t\right)=u_2^{mpl}\left(x_{22},y_{22},t\right)-z_{22}\left(x\right)\dfrac{\partial u_3^{mp}}{\partial x} \\[3mm] \quad\quad u_3^{22}\left(x_{22},y_{22},z_{22},t\right)=u_3^{mp}\left(x,y,t\right) \end{array} \right\} \tag{9.4}$$

The health plate part 3 and 4 displacement may be calculated by means of the subsequent Equation 9.5:

$$
\left\{
\begin{aligned}
u_1(x,y,z,t) &= u_1^{mpl}(x,y,t) - z(x)\frac{\partial u_3^{mp}}{\partial x} \\
u_2(x,y,z,t) &= u_2^{mpl}(x,y,t) - z(x)\frac{\partial u_3^{mp}}{\partial y} \\
u_3(x,y,z,t) &= u_3^{mp}(x,y,t)
\end{aligned}
\right\}
\tag{9.5}
$$

9.4.1 STRAIN ENERGY FORMULATION

Strain energy expression is constructed utilising extension coupling and stiffness matrices related to bending in a laminated composite plate. The displacement field is carefully examined on the basis of classic laminated plate theory (CLPT). The linear deformations in the x and y axes determines the summation of all stress energy; the plate transverse deflection from the centre is $\{U_{mi}\}$ and $\{U_{md}\}$ (Arun Kumar & Mallikarjuna Reddy, 2020).

Where, i = intact plate section and d = unhealth section, linear strain energy may be constructed by assuming a displacement field (as shown in Equations 9.3, 9.4 and 9.5)due to un-delaminated and delaminated composite plate section bending and membrane displacement.

$$
U_{mi} = \frac{1}{2} \int_0^l \int_{-\frac{w}{2}}^{\frac{w}{2}}
\left\{
\begin{array}{c}
\frac{\partial u_1^{mp}}{\partial x} \\[6pt]
\frac{\partial u_2^{mp}}{\partial y} \\[6pt]
\frac{\partial u_1^{mp}}{\partial y} + \frac{\partial u_2^{mp}}{\partial x} \\[6pt]
-\frac{\partial^2 u_3^{mp}}{\partial x^2} \\[6pt]
-\frac{\partial^2 u_3^{mp}}{\partial y^2} \\[6pt]
-2\frac{\partial^2 u_3^{mp}}{\partial x \partial y}
\end{array}
\right\}
\times
\begin{bmatrix}
A_{ij}^{es} & B_{ij}^{es} \\
B_{ij}^{es} & D_{ij}^{bs}
\end{bmatrix}
\times
\left\{
\begin{array}{c}
\frac{\partial u_1^{mp}}{\partial x} \\[6pt]
\frac{\partial u_2^{mp}}{\partial y} \\[6pt]
\frac{\partial u_1^{mp}}{\partial y} + \frac{\partial u_2^{mp}}{\partial x} \\[6pt]
-\frac{\partial^2 u_3^{mp}}{\partial x^2} \\[6pt]
-\frac{\partial^2 u_3^{mp}}{\partial y^2} \\[6pt]
-2\frac{\partial^2 u_3^{mp}}{\partial x \partial y}
\end{array}
\right\}
dxdy
\tag{9.6}
$$

9.4.2 FORMULATION OF TRANSFORMATION MATRIX

$$
[T_\alpha] =
\begin{bmatrix}
l^2 & m^2 & -2\,lm \\
m^2 & l^2 & 2\,lm \\
lm & -lml^2 & -m^2
\end{bmatrix}
\tag{9.7}
$$

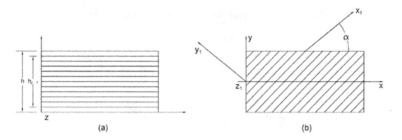

FIGURE 9.2   (a) Ply-thickness representation used for determining stiffness of the plate and (b) coordinates conversion from (x_l, y_l, z_l) to (x, y, z).

Where, $l = cos\alpha$ and $m = sin\alpha$, $[T_\alpha]$ is the transformation of stresses; matrix (x_l, y_l, z_l) from the main plate axis (x, y, z) is represented by the coordinate system. The coordinate indicates the ply orientation (x_l, y_l, z_l) whereas the global co-ordinates derived from the local ply axis display (x_l, y_l, z_l). Figure 9.2 illustrates the laminated plate orientation and ply thickness.

9.4.3   Finite Element Approach

Two requirements, the number of nodes and degrees of freedom, are fundamental to formulating the finite element. 5 DOF per node in the rectangle element are modelled.

The 5 DOF include the displacements: $u_1^{mp}, u_2^{mp}, u_3^{mp}, \theta_{xx}$ and θ_{yy} where u_1^{mp} and u_2^{mp} are the direction displacements in x- and y-axis of a composite plate respectively, and whereas u_3^{00} is the transverse deflection of the composite structure as well as $\dfrac{\partial u_3^{mp}}{\partial y} = \theta_{xx}$ and $\dfrac{\partial u_3^{mp}}{\partial y} = -\theta_{yy}$ are the changes accordingly caused by the x- and y-axis rotations of the composite structures. Therefore the displacement may be illustrated briefly as indicated in Equation 9.8 related to the deformation in the finite element plate at any point.

$$\{u^1\} = \left[N_i^1(x,y)\right]\{d_1\} \tag{9.8}$$

Where i – total number of nodes = 1, 2 ... 4 & N^1_i - nodal shape function. The strain – displacement matrix relation for the rectangular element is given in Equation 9.9:

$$\{\chi_I\} = \left[\bar{B}_I\right]\{u^1\} \tag{9.9}$$

Where $\{\chi_I\}$— vector of strain, d_I– vector of displacement, $\{u^1\}$ – point of elemental displacement and $\left[\bar{B}_I\right]$ – element strain – displacement matrix (Arumugam Ananda Babu & Vasudevan, 2017; A. Ananda Babu et al., 2016; Hirwani et al., 2018; and) Arun Kumar & Mallikarjuna Reddy, 2020).

9.4.4 FORMULATION OF ELEMENT STIFFNESS AND MASS MATRICES

Previously Equations 9.8–9.9 focused on the strain-kinetics energy equations, the mass and stiffness matrices of four-noded element. The composite plate delamination is expressed in form of equations of motion as given in Equation 9.10.

$$\left[m^{pp} \right]\left\{ \ddot{d}_l \right\} + \left[k \right]\left\{ d_l \right\} = \left\{ f^{pp} \right\} \tag{9.10}$$

$[m^{pp}],[k^{pp}], \{d^l\}$ & $\{f^{pp}\}$ are local mass, local stiffness, local vector of displacement at each node and local force at each node, respectively. The composite plate delamination is expressed in form of equations of motion as given in Equation 9.11.

$$\left[\bar{M} \right]\left\{ \ddot{u} \right\} + \left[\bar{K} \right]\left\{ u \right\} = \left\{ \bar{F} \right\} \tag{9.11}$$

$\left[\bar{M} \right], \left[\bar{K} \right]$ and $\left\{ \bar{F} \right\}$ are the global mass, global stiffness and global force vector, respectively (Arun Kumar & Mallikarjuna Reddy, 2020).

9.5 CONCLUSION

Composite materials continue to be a commonly used alternative for aircraft parts and structure building. However, due to the increased usage and prominence, how these materials function under different load conditions must be taken into account, especially when they relate to their properties, the related failure modes, and the analyses of the broken material when affected. The related works regarding failure and fracture analysis of composite material has been reviewed to explain the present state of the science. Modelling of delamination cracks in the composite plate structure using finite element method is also discussed.

REFERENCES

Ananda Babu, A., Edwin Sudhagar, P., & Rajamohan, V. (2016). Dynamic characterization of thickness tapered laminated composite plates. JVC/Journal of Vibration and Control, 22(16), 3555–3575. doi:10.1177/1077546314564588

Arun Kumar, K., & Mallikarjuna Reddy, D. (2020). Different interface delamination effects on laminated composite plate structure under free vibration analysis based on classical laminated plate theory. Smart Materials and Structures, 29(11), 115028. doi:10.1088/1361-665X/abaec7

Babu, A.A. and Vasudevan, R. (2017). Vibration analysis of rotating delaminated non-uniform composite plates. Aerospace Science and Technology, 60(November), 172–182. doi:10.1016/j.ast.2016.11.009

Belingardi, G., Martella, P., & Peroni, L. (2007). Fatigue analysis of honeycomb-composite sandwich beams. Composites Part A: Applied Science and Manufacturing, 38(4), 1183–1191. doi:10.1016/j.compositesa.2006.06.007

Daniel, I.M., Gdoutos, E.E., Wang, K.A. and Abot, J.L. (2002). Failure modes of composite sandwich beams. International Journal of Damage Mechanics, 11(4), 309–334.

Gdoutos, E. E., Daniel, I. M., & Wang, K.-A. (2003). Compression facing wrinkling of composite sandwich structures. Mechanics of Materials, 35(3–6), 511–522.

Gopal, K. V. N. (2016). Product design for advanced composite materials in aerospace engineering. In Advanced composite materials for aerospace engineering (pp. 413–428). Elsevier.

Hiken, A. (2018). The evolution of the composite fuselage: A manufacturing perspective. In Aerospace Engineering. IntechOpen.

Hirwani, C. K., Panda, S. K., Mahapatra, S. S., Mandal, S. K., Srivastava, L., & Buragohain, M. K. (2018). Flexural strength of delaminated composite plate – An experimental validation. International Journal of Damage Mechanics, 27(2), 296–329. doi:10.1177/1056789516676515

Kassapoglou, C. (2013). Design and analysis of composite structures: with applications to aerospace structures. John Wiley & Sons.

Mostafa, N. H., Ismarrubie, Z. N., Sapuan, S. M., & Sultan, M. T. H. (2016). Effect of fabric biaxial prestress on the fatigue of woven E-glass/polyester composites. Materials & Design, 92, 579–589.

Opelt, C. V., Cândido, G. M., & Rezende, M. C. (2018). Compressive failure of fiber reinforced polymer composites--A fractographic study of the compression failure modes. Materials Today Communications, 15, 218–227.

Peters, S. T. (1998). Handbook of composites, Champan & Hall. London, 839, 855.

Shikhmanter, L., Eldror, I., & Cina, B. (1989). Fractography of unidirectional CFRP composites. Journal of Materials Science, 24(1), 167–172.

Soutis, C. (2005a). Fibre reinforced composites in aircraft construction. In Progress in aerospace sciences (Vol. 41(2), pp. 143–151). Elsevier Ltd. doi:10.1016/j.paerosci.2005.02.004

Soutis, C. (2005b). Carbon fiber reinforced plastics in aircraft construction. Materials Science and Engineering A, 412(1–2), 171–176. doi:10.1016/j.msea.2005.08.064

Zimmermann, N., & Wang, P. H. (2020). A review of failure modes and fracture analysis of aircraft composite materials. Engineering Failure Analysis, 115(April), 104692. doi:10.1016/j.engfailanal.2020.104692

10 Low Temperature Effect on Deflection Behavior of a Cracked Laminated Glass Plate

Tayfun Inik and Mehmet Yetmez

Zonguldak Bulent Ecevit University, Zonguldak, Turkey

CONTENTS

10.1 INTRODUCTION

In last four decades, the use of glass in buildings has been growing significantly. It is obvious that, at first, monolithic glass was used as a window component. Later, as a modern design, it was constantly being used for a load-carrying element of roofs, beams, columns and floors (Belis et al., 2011). Now, the ongoing demand for such technology has required the use of further complications to the design process. Consequently, it is clearly seen that the development of the glass industry needs better structural design processes and better understanding of the load-carrying capacity for improving durability and service life of glass.

Monolithic glass is a well-known brittle material with large scatter in its fracture strength. Due to the existence of microscopic surface flaws and their geometrical properties, laminated glass has been more preferable than monolithic glass since the past four decades (Osnes et al., 2020). On one hand, in the beginning, PVB (polyvinyl butyral) was not considered as a major effect of nonlinear behavior of laminated glass under mechanical conditions. PVB is usually characterized as linear-viscoelastic, i.e.

DOI: 10.1201/9781003200994-10

its mechanical properties are dependent of the load application time and the temperature working conditions (Nicholas W. Tschoegl, 1989). Furthermore, in the glass community it is said that the mechanical properties of PVB change after the process in autoclave. This means that the PVB exhibits different behavior when it is tested freely (not constrained in a laminate) or when it is tested in constrained conditions in laminated glass elements (López-Aenlle et al., 2019). Particularly, the effect of PVB thickness as a major layer or number of layers plays an important role in critical deflection and buckling force capacity (Shaterzadeh et al., 2014). On the other hand, later on, it has been clearly seen that PVB is a very important part of the laminated glass unit on which to compute failure stress analysis (B. R. A. Behr et al., 1985). In addition to its mechanical condition, variation of temperature influences stress behavior of laminated glass plate, especially post-crack response characteristics (Samieian et al., 2018). Despite some advancements, more progress is still required to understand temperature effects on strain field behavior and fundamental vibration characteristics of cracked laminated plates (Centelles et al., 2019; López-Aenlle & Pelayo, 2019). First, the usage of laminated glass in structural design process is quite common today. Because of its nonlinear behavior depending on geometry and material, numerical and experimental studies become important to investigate the PVB effect for laminated glasses (López-Aenlle et al., 2019). Basically, temperature and interlayer thickness of the layered glass components possess a key role for the nonlinear approaches (Boutin et al., 2021; Galuppi & Royer-Carfagni, 2012; Aşik & Tezcan, 2005; Hooper et al., 2012; Chen et al., 2013). Second, crack behavior of laminated glass is commonly considered by finite element method (FEM) to find fracture toughness (K_1, K_2 and J-Integral (JINT)) (Tsai & Chen, 2005; Shokrieh & Zeinedini, 2014; X. Chen & Chan, 2018). Third, natural frequency and damping characteristics of the laminated glass are also taken to analyse the accuracy of the response of effective thickness and other simplified approaches (Zemanová et al., 2018; Koutsawa & Daya, 2007).

The aim of this chapter is to investigate low temperature effects on deflection behavior of a cracked laminated circular glass plate, both experimentally and numerically. The circular laminated plates are handled with two groups: no crack and a surface crack. Effect of PVB thickness is also examined under different loading and thermal conditions. Additionally, free vibration characteristics of the laminated glass circular plates are considered in order to understand the deflection behavior clearly.

10.2 MATERIALS AND METHODS

As a laminated safety glass, laminated glass specimens used in this study consist of monolithic glass and PVB layers (R. A. Behr et al., 1993). The layers are bonded together mechanically and chemically through a process including heat and pressure in an autoclave (Osnes et al., 2020; Vedrtnam & Pawar, 2017; Foraboschi, 2012). In this study, laminated glass specimens provided by Sisecam (Sisecam A.Ş., Turkey) are three-layered circular plates with 160 mm in diameter and 3 mm in monolithic thickness. Three thickness values of a PVB interlayer are 0.38, 0.76 and 1.52 mm. Details for the laminated glass specimens are given in Table 10.1.

It is obvious that glass plates primarily fail in tension, and the fracture strength is dependent on the applied stresses and the properties of the surface flaws

TABLE 10.1
Properties of Materials Used in Laminated Glass

Layer	ρ (g/cm³)	E (GPa)	ν (mm/mm)	α (×10⁻⁶, 1/°C)	Reference
	2.35	20	0.22	9	(Xu et al., 2011)
Glass					(Vedrtnam & Pawar, 2017)
PVB	1	0.7	0.44	15.1	(Vedrtnam & Pawar, 2017)

(Osnes et al., 2018; Timoshenko & Woinowsky-Krieger, 1959; Beason & Morgan, 1984). Consequently, in addition to PVB thickness, second testing parameter of this study is central surface crack with 10 mm in length and 0.1 mm in depth. Furthermore, low temperature and pressure applied to the bottom-layer surface of a laminated circular glass plate are the last parameters. On one hand, depending on these conditions, the effects of temperature, large deflection and surface crack on behavior of a laminated circular glass plate are investigated. On the other hand, due to the toughness and flexibility properties of PVB (Osnes et al., 2020; Zhang et al., 2015), free vibration behavior of a cracked laminated glass circular plates are also considered to understand the deflection behavior clearly.

10.2.1 QUASI-STATIC TESTING

Experimental setup with liquid nitrogen tank, LVDT and data collection system are given in Figure 10.1.

Each specimen is insulated with a 2 mm-thick head cylinder gasket. The specimen is properly connected to the experimental set-up and positioned. After centering, to connect the lower clamping plate with upper clamping plate, eight equidistant M16 bolts are used to connect the clamping plates together with a torque of 120 Nm. The bolt length is 85 mm. In order to get the low temperature, a liquid nitrogen tank is connected with inlet cryogenic valve flex hose. The temperature range considered

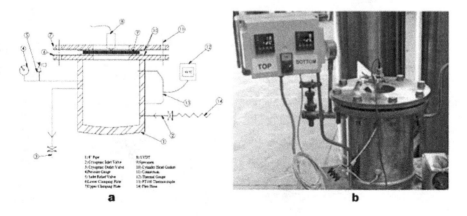

FIGURE 10.1   Representation of the experimental setup: (a) details, (b) general view.

in this study is between 0 to $-100°C$. Additionally, applied pressure range varies from 0 to 1 MPa.

10.2.2   MODAL ANALYSIS

Experimental free vibration analysis is considered by the following steps: (i) an impulse hammer with a force transducer is used to excite each of the uncracked/cracked circular laminated plates through the selected point; (ii) the responses are obtained by an accelerometer; and (iii) the measurements are completed using a microprocessor-based data acquisition system, namely SoMat™ eDAQ-lite and nCode GlyphWorks software (HBM, Inc., USA).

10.2.3   NUMERICAL MODEL

General purpose finite element code Ansys (Model No: 19.2, 2018 Ansys, Inc., USA) is taken into account for the numerical model and its analysis. Numerical analysis is considered by the following steps:

(i)   Static structure is used to find directional deformation strain energy, K_1, K_2 and J-integral (JINT) for uncracked/cracked circular laminated plates. Both SOLID186 and SOLID187 are preferred in this part of the analysis (#nodes: 71288 and #elements: 54362) (see Figure 10.2).

(ii) Modal analysis is conducted to find the natural frequencies and damping ratios vwith element type SOLID187 (#nodes: 62525 and #elements 30482) (see Figure 10.3).

10.3   RESULTS AND DISCUSSION

Experimental and numerical results of variations of major normal stress (σ_z) and deflection (Δ_z) for a laminated glass with no crack and with a central surface crack are presented in Figures 10.4–10.6. It can be noted that numerical results are

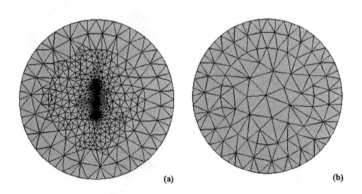

(a)   (b)

FIGURE 10.2   Static structural mesh model: (a) with a surface crack, (b) with no crack.

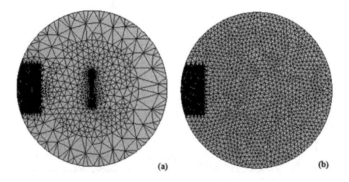

FIGURE 10.3 Modal analysis mesh model: (a) with a crack, (b) with no crack.

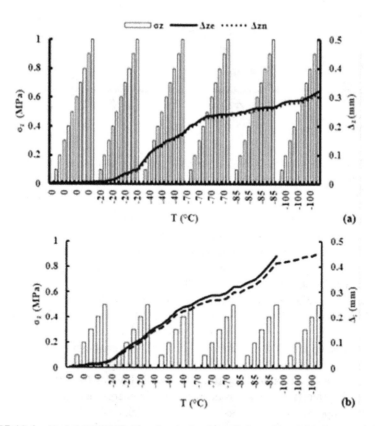

FIGURE 10.4 Variation of deflection for the laminated glass plates with 3 mm + 1.52 mm + 3 mm: (a) the plate with no crack, (b) the plate with a crack (experimental results: Δze, numerical results: Δzn).

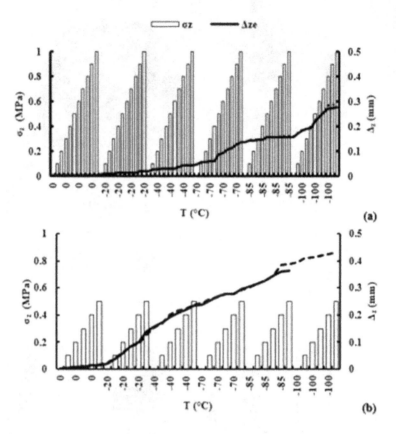

FIGURE 10.5 Variation of deflection for laminated glass with 3 mm + 0.76 mm + 3 mm: (a) the plate with no crack, (b) the plate with a crack (experimental results: Δze, numerical results: Δzn).

compatible with the experimental evaluations. Figures 10.4–10.6 indicate that (i) characteristic stress and deflection responses for 0.38 mm start slightly while responses for 0.76 mm begin moderately at −20°C, (ii) decreasing temperature effect for 1.52 mm may be observed clearly at −20°C, (iii) In the range of −85 to −100°C, there are sharp rising values of major normal stress (σ_z) and deflection (Δ_z) for both interlayer thicknesses of 0.38 and 0.76 mm, (iv) for cracked plate with 1.52 mm interlayer thickness, decreasing temperature gives approximately linear variations for the stress and deflection up until a temperature of −85°C as fracture point.

Numerical results of variation of strain energy through diameter of cracked laminated glass plates at room temperature are given in Figures 10.7–10.9. It is seen that (i) decreasing interlayer thickness decreases strain energy, and (ii) sharp increase of fluctuations appears when decreasing interlayer thickness.

Numerical results of variation of shear stress through diameter of cracked laminated glass plates at room temperature are given in Figures 10.10–10.12. It is concluded that (i) in the same way as strain energy, decreasing interlayer thickness

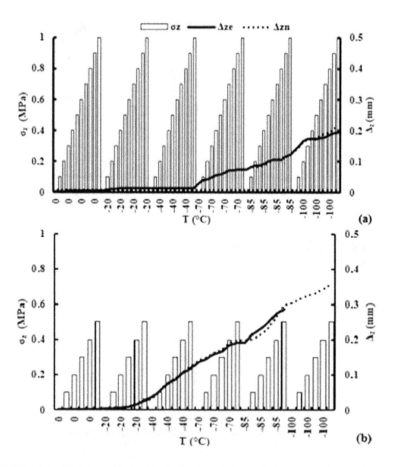

FIGURE 10.6 Variation of deflection for laminated glass with 3 mm + 0.38 mm + 3 mm: (a) the plate with no crack, (b) the plate with a crack (experimental results: Δze, numerical results: Δzn).

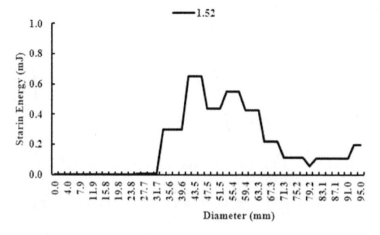

FIGURE 10.7 Variation of strain energy through diameter of cracked laminated glass plate with 3 mm + 1.52 mm + 3 mm.

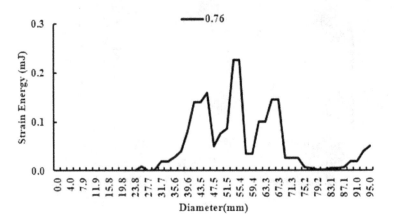

FIGURE 10.8   Variation of strain energy through diameter of cracked laminated glass plate with 3 mm + 0.76 mm + 3 mm.

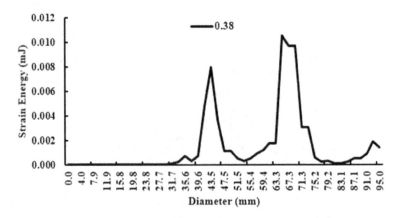

FIGURE 10.9   Variation of strain energy through diameter of cracked laminated glass plate with 3 mm + 0.38 mm + 3 mm.

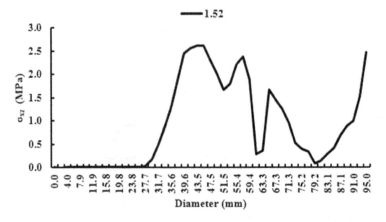

FIGURE 10.10   Variation of shear stress through diameter of cracked laminated glass plate with 3 mm + 1.52 mm + 3 mm.

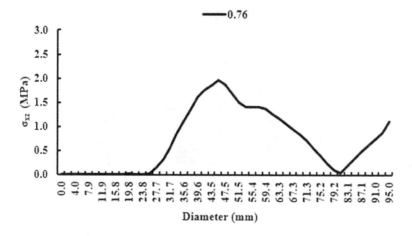

FIGURE 10.11 Variation of shear stress through diameter of cracked laminated glass plate with 3 mm + 0.76 mm + 3 mm.

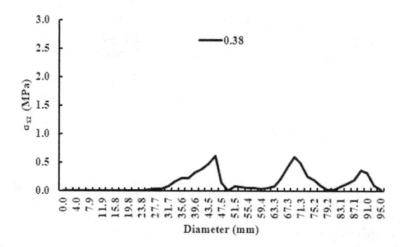

FIGURE 10.12 Variation of shear stress through diameter of cracked laminated glass plate with 3 mm + 0.38 mm + 3 mm.

TABLE 10.2
Stress Intensity Factors and JINT Values of a Cracked Laminated Glass Plate

PVB thickness (mm)	K_1 (MPa $\sqrt{mm}$)	K_2 (MPa $\sqrt{mm}$)	JINT (MPa mm)
1.52	38.075	21.775	0.084
0.76	43.421	13.728	0.074
0.38	40.623	1.4671	0.078

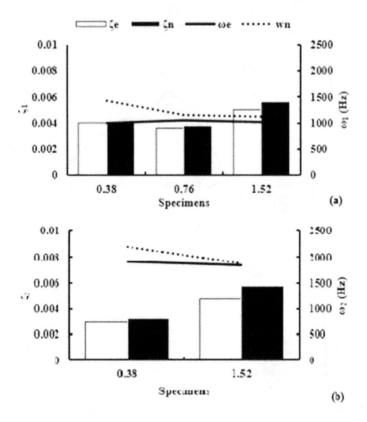

FIGURE 10.13 Modal analysis results of a laminated glass plate before low temperature testing: (a) mode-1, (b) mode-2 (experimental results: ωe, ζe, numerical results: ωn, ζn).

decreases shear stress, (ii) centre through shear edge effect decreases when decreasing interlayer thickness (see K_2 values in Table 10.2).

Modal analysis results of a laminated glass plate before and after low temperature testing are given in Figures 10.13 and 10.14. One may conclude that (i) finite element model for the dynamic analysis is slightly acceptable, (ii) after low temperature testing, both natural frequency and damping ratio values increase slightly for interlayer thickness of 1.52 mm, and (iii) crack effect on free vibration characteristics are clearly observed for all three interlayer thicknesses.

10.4   CONCLUSION

Mechanical characteristics of a laminated glass plate at low temperatures are investigated by considering three main interlayer thicknesses. External stress and temperature are applied to the lower monolithic glass layer. The effect of stress and temperature variations on a central surface crack at the upper monolithic glass layer is also taken into account. Experimental and numerical results indicate that (i) there are nonlinear displacement and stress behaviors through the lower temperatures up to −100°C, (ii)

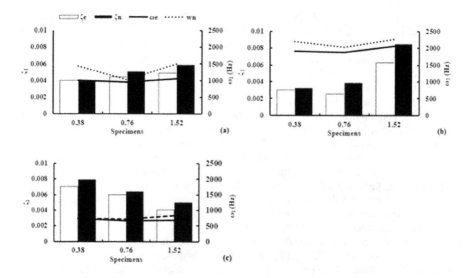

FIGURE 10.14 After low temperature testing, modal results of (a) mode-1 of a laminated glass plate with no crack, (b) mode-2 of a laminated glass plate with no crack (c) mode-1 of a laminated glass plate with a crack (experimental results: ω_e, ζ_e, numerical results: ω_n, ζ_n).

PVB thickness is the key for critical mechanical characteristics such as elastic strain energy and shear stress, (iii) temperature range between −50 and −85°C is critical for the fracture mechanics point of view, and (iv) fracture pattern in temperature range of −50 and −85°C is to be focused on for the interlayer studies.

ACKNOWLEDGEMENT

The authors gratefully acknowledge Mr. Resit Kacar (Şişecam A.Ş., Tuzla Turkey) for his help of providing laminated glass specimens; Mr. Osman Cihan and Mr. Ayhan Yurdabakan (May Otomasyon Gaz Sistemleri Ltd., Tuzla Turkey) for their help with the experimental arrangement.

REFERENCES

Ansys Workbench, Model No: 19.2, ©1971-2018, Ansys, Inc., USA
Aşik, M. Z., & Tezcan, S. (2005). A mathematical model for the behavior of laminated glass beams. *Computers and Structures*, 83(21–22), 1742–1753. doi:10.1016/j.compstruc.2005.02.020
Beason, W. L., & Morgan, J. R. (1984). Glass failure prediction model. *Journal of Structural Engineering*, 110(2), 197–212. doi:10.1061/(asce)0733-9445(1984)110:2(197)
Behr, B. R. A., Minor, J. E., Linden, M. P., & Vallabhan, C. V. P. (1985). Laminated glass units under uniform pressure. *Journal of Structural Engineering*, 111(5), 1037–1050. doi:10.1061/(asce)0733-9445(1985)111:5(1037)
Behr, R. A., Minor, J. E., & Norville, H. S. (1993). Structural behavior of architectural laminated glass. *Journal of Structural Engineering*, 119(1), 202–222. doi:10.1061/(asce)0733-9445(1993)111:1(202)

Belis, J., Mocibob, D., Luible, A., & Vandebroek, M. (2011). On the size and shape of initial out-of-plane curvatures in structural glass components. *Construction and Building Materials*, 25(5), 2700–2712. doi:10.1016/j.conbuildmat.2010.12.021

Boutin, C., Viverge, K., & Hans, S. (2021). Dynamics of contrasted stratified elastic and viscoelastic plates - application to laminated glass. *Composites Part B: Engineering*, 212, 108551. doi:10.1016/j.compositesb.2020.108551

Centelles, X., Castro, J. R., & Cabeza, L. F. (2019). Experimental results of mechanical, adhesive, and laminated connections for laminated glass elements – A review. *Engineering Structures*, 180(April 2018), 192–204. doi:10.1016/j.engstruct.2018.11.029

Chen, X., & Chan, A. H. C. (2018). Modelling impact fracture and fragmentation of laminated glass using the combined finite-discrete element method. *International Journal of Impact Engineering*, 112(April 2017), 15–29. doi:10.1016/j.ijimpeng.2017.10.007

Chen, J., Xu, J., Yao, X., Liu, B., Xu, X., Zhang, Y., & Li, Y. (2013). Experimental investigation on the radial and circular crack propagation of PVB laminated glass subject to dynamic out-of-plane loading. *Engineering Fracture Mechanics*, 112–113, 26–40. doi:10.1016/j.engfracmech.2013.09.010

Foraboschi, P. (2012). Analytical model for laminated-glass plate. *Composites Part B: Engineering*, 43(5), 2094–2106. doi:10.1016/j.compositesb.2012.03.010

Galuppi, L., & Royer-Carfagni, G. F. (2012). Effective thickness of laminated glass beams: New expression via a variational approach. *Engineering Structures*, 38, 53–67. doi:10.1016/j.engstruct.2011.12.039

Hooper, P. A., Blackman, B. R. K., & Dear, J. P. (2012). The mechanical behaviour of poly(vinyl butyral) at different strain magnitudes and strain rates. *Journal of Materials Science*, 47(8), 3564–3576. doi:10.1007/s10853-011-6202-4

Koutsawa, Y., & Daya, E. M. (2007). Static and free vibration analysis of laminated glass beam on viscoelastic supports. *International Journal of Solids and Structures*, 44(25–26), 8735–8750. doi:10.1016/j.ijsolstr.2007.07.009

López-Aenlle, M., Noriega, A., & Pelayo, F. (2019). Mechanical characterization of polyvinil butyral from static and modal tests on laminated glass beams. *Composites Part B: Engineering*, 169(February), 9–18. doi:10.1016/j.compositesb.2019.03.077

López-Aenlle, M., & Pelayo, F. (2019). Static and dynamic effective thickness in five-layered glass plates. *Composite Structures*, 212(July 2018), 259–270. doi:10.1007/978-3-642-73602-5

Osnes, K., Børvik, T., & Hopperstad, O. S. (2018). Testing and modelling of annealed float glass under quasi-static and dynamic loading. *Engineering Fracture Mechanics*, 201(7491), 107–129. doi:10.1016/j.engfracmech.2018.05.031

Osnes, K., Hopperstad, O. S., & Børvik, T. (2020). Rate dependent fracture of monolithic and laminated glass: Experiments and simulations. *Engineering Structures*, 212(February), 110516. doi:10.1016/j.engstruct.2020.110516

Samieian, M. A., Cormie, D., Smith, D., Wholey, W., Blackman, B. R. K., Dear, J. P., & Hooper, P. A. (2018). Temperature effects on laminated glass at high rate. *International Journal of Impact Engineering*, 111, 177–186. doi:10.1016/j.ijimpeng.2017.09.001

Shaterzadeh, A. R., Abolghasemi, S., & Rezaei, R. (2014). Finite element analysis of thermal buckling of rectangular laminated composite plates with circular cut-out. *Journal of Thermal Stresses*, 37(5), 604–623. doi:10.1080/01495739.2014.885322

Shokrieh, M. M., & Zeinedini, A. (2014). A novel method for calculation of strain energy release rate of asymmetric double cantilever laminated composite beams. *Applied Composite Materials*, 21(3), 399–415. doi:10.1007/s10443-013-9328-5

Timoshenko, S., & Woinowsky-Krieger, S. (1959). *Theory of plates and shells*. 2nd Edition. Mc Graw-Hill, New York.

Tsai, G. C., & Chen, J. W. (2005). Effect of stitching on Mode I strain energy release rate. *Composite Structures*, 69(1), 1–9. doi:10.1016/j.compstruct.2004.02.009

Tschoegl, N.W. (1989). *The Phenomenological Theory of Linear Viscoelastic Behavior.* Springer, Berlin, Heidelberg. doi:10.1016/j.compstruct.2019.01.037

Vedrtnam, A., & Pawar, S. J. (2017). Laminated plate theories and fracture of laminated glass plate – A review. *Engineering Fracture Mechanics*, 186, 316–330. doi:10.1016/j.engfracmech.2017.10.020

Xu, J., Li, Y., Liu, B., Zhu, M., & Ge, D. (2011). Experimental study on mechanical behavior of PVB laminated glass under quasi-static and dynamic loadings. *Composites Part B: Engineering*, 42(2), 302–308. doi:10.1016/j.compositesb.2010.10.009

Zemanová, A., Zeman, J., Janda, T., Schmidt, J., & Šejnoha, M. (2018). On modal analysis of laminated glass: Usability of simplified methods and Enhanced Effective Thickness. *Composites Part B: Engineering*, 151, 92–105. doi:10.1016/j.compositesb.2018.05.032

Zhang, X., Hao, H., Shi, Y., & Cui, J. (2015). The mechanical properties of Polyvinyl Butyral (PVB) at high strain rates. *Construction and Building Materials*, 93, 404–415. doi:10.1016/j.conbuildmat.2015.04.057

11 Damage Detection and Evaluation on Gamma Irradiated Oil Palm EFB/Kevlar Hybrid Composites Using NDT Methods

Siti Madiha Muhammad Amir

Universiti Putra Malaysia, Serdang, Malaysia

Malaysian Nuclear Agency, Bangi, Malaysia

Mohamed Thariq Hameed Sultan

Universiti Putra Malaysia, Serdang, Malaysia

Institute of Tropical Forestry and Forest Products (INTROP), Universiti Putra Malaysia, Serdang, Malaysia

Aerospace Malaysia Innovation Centre (944751-A), Cyberjaya, Malaysia

Mohammad Jawaid

Institute of Tropical Forestry and Forest Products (INTROP), Universiti Putra Malaysia, Serdang, Malaysia

Mohamad Ridzuan Ahmad

Malaysian Nuclear Agency, Bangi, Malaysia

Muhamad Noor Izwan Ishak

Malaysian Nuclear Agency, Bangi, Malaysia

Suhairy Sani

Malaysian Nuclear Agency, Bangi, Malaysia

DOI: 10.1201/9781003200994-11

Syafiqah Nur Azrie Safri

Institute of Tropical Forestry and Forest Products (INTROP),
Universiti Putra Malaysia, Serdang, Malaysia

Ain Umaira Md Shah

Universiti Putra Malaysia, Serdang, Malaysia

Shukri Mohd and Khairul Anuar Mohd Salleh

Malaysian Nuclear Agency, Bangi, Malaysia

CONTENTS

11.1   INTRODUCTION

Damages such as matrix cracking, delamination, fibre breakage, perforation of fibre-matrix surface in laminated polymer matrix composite, fibre-matrix debonding and fibre pull-out commonly occur during low velocity impact events [1, 2]. Detecting the impact damage and its failure modes from the low velocity impact event is important because this is related to the composite's residual strength. Furthermore, knowing the interaction between the failure modes provides the understanding of the damage mode initiation and propagation [3].

Sivakumar et al. [4] studied the low-velocity impact response of short oil palm EFB fibre–reinforced metal laminates using Charpy test on flatwise and edgewise of the composites. It was found that the flatwise impact properties displayed better impact resistance compared to edgewise impact properties. Zainudin et al. [5] studied the impact strength of the hybrid composites between coir/oil palm empty fruit bunch. The impact studies were conducted using the Izod pendulum method with energy of 2.75J. Various combinations of weight ratio were fabricated and from the results it was observed that composites made from 30%wt coir, 30%wt oil palm EFB and 70%wt polypropylene displayed the highest impact resistance.

Composites made from Kevlar were also commonly being investigated using low-velocity impact testing. Channabasavaraju [6] studied the impact response of laminated Kevlar composites with different impact energy. From the work, it was observed that the impact damage developed around the point of the impact which resulted in loss of strength. The impact force caused the local delamination of the composite skin. Taraghi et al. [1] in his work investigated the low-velocity impact response of woven Kevlar/epoxy laminated composites with addition of multi-walled carbon nanotubes at temperatures of 27⁰C and −40⁰C. The impact tests were performed at different energy levels from 20J to 80J. From the work, it was observed that by adding the multi-walled carbon nanotubes, the size of damage decreased and the delamination around the impact point diminished.

Low velocity impact damage studies had also been done on Kevlar hybrid composites. Investigation on low-velocity impact response on carbon-aramid/epoxy hybrid composites has been conducted [7]. From the study, the results show that carbon fabric laminates on the impact surface show a smaller damage area than aramid fabric laminates. Gustin et al. [8] evaluated the low velocity impact response of sandwich composites consisting of Kevlar/carbon fibre. It was found that the maximum absorbed energy is improved with the addition of Kevlar to the face sheet. The impact resistance properties of Kevlar/glass fibre hybrid composite laminates were also studied by Shaari et al. [9]. The authors concluded that adding Kevlar fibre to glass fibre has improved the damage area, energy absorbed and load carrying capability. Hence, Kevlar has better resistance towards impact loading due to its ability to absorb more energy. This contributes to smaller damage size in Kevlar as compared to glass fibre. Imielinska et al. [10] studied low-velocity impact behaviour on water immersion ageing woven aramid-glass fibre/epoxy composites and found out that the impact energy absorption was not significantly affected by the absorbed water. Another study on the low-velocity impact response of hybrid kenaf-Kevlar composites showed that different types of fibre combinations produced different results [11]. Besides, the thickness of the laminates affected the impact dynamics of the hybrid composites.

Because low-velocity impact damage causes internal damage to the structure with very minimal visual detectability, non-destructive testing (NDT) is needed to detect the internal damage. Since hybrid composites between synthetic and natural fibres have gained wide interest, conventional NDT methods are not able to perform an inspection of the hybrid composites, so advanced NDT techniques or combination NDT methods are required instead.

Fidan et al. [12] visualise the internal damage impact using micro-computerised tomography (micro-CT) on glass-fibre–reinforced and glass fibre with aramid fibre–reinforced polyester composites after being impacted at 80J energy. From the investigation, the results show that the 3D delamination pattern defect in glass-reinforced composites is more visible due to the nature of the glass fibre. However, the delamination pattern lost its effectiveness when aramid fibre was added to the glass fibre.

Other than that, the failure process and characteristics of carbon/Kevlar hybrid woven composites under a high strain-rate impact have also been investigated [13].

The acoustic emission (AE) technique was used as the non-destructive method in this work. The parameters of AE signal cumulative counts and amplitude provide the information on plastic deformation and fibrillation of fibres, matrix cracking propagation and fibre breakages. The structural changes in a stressed material could be obtained from the technique. Jaroslac et al. [14] applied the AE technique on a tensile test on composite materials reinforced with carbon and aramid fibres. Information regarding the separation of the matrix, the extraction of fibre from the matrix and the breaking of the fibres was obtained from the experiments. The Root mean square (RMS) was observed in this work.

The liquid crystal thermography method was applied to inspect delamination and air bubbles in a hybrid of Kevlar/resin as the skin and glass/resin as the core [15]. Destic et al. [16] conducted NDT using the THz imaging set-up on Kevlar fibres. The results obtained were promising, whereby delamination in Kevlar and a break in a carbon/epoxy samples were detected.

Based on the literature studies, it was found that until this date, none of the researchers has studied the impact-damage properties of Kevlar/oil palm hybrid composites. It was also found that the researchers are focusing more on the non-destructive testing on hybrid composites made from only synthetic/synthetic fibres. Limited studies were done on the hybrid composites made from synthetic/natural fibres using NDT methods. Hence, this research focus on the evaluation of low-velocity impact damage on hybrid composites made from oil palm EFB and Kevlar. In this work, ultrasonic, computed tomography and acoustic emission techniques were applied to evaluate the impact damage.

11.2 MATERIALS AND METHODS

In this research, hybrid composites were made from Kevlar and oil palm EFB. All samples were prepared using the traditional hand lay-up method. The stacking sequence of the hybrid composites fabricated is shown in Figure 11.1. After the fabrication process, the samples were cut into dimensions of 150 mm × 200 mm and sent for gamma irradiation with different radiation doses of 0kGy, 25 kGy, 50 kGy and 150 kGy.

Low-velocity impact test was performed on the hybrid composites using drop test rig model IMATEK IM10. A striker with weight of 5.101 kg was dropped from several heights with gravitational constant 9.82 ms^{-2}. Various energies were applied to the hybrid composites as tabulated in Table 11.1.

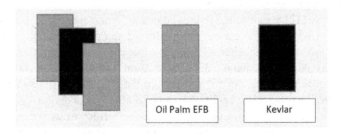

FIGURE 11.1 Stacking sequence of the hybrid composites.

TABLE 11.1

Energy Level for Low Velocity Impact Event

Energy level, J	Height, m
10	0.2
15	0.3
20	0.4
25	0.5
30	0.6
35	0.7
40	0.8

In this work, the impact damage was analysed using X-ray computed tomography system as shown in Figure 11.2. The unit is equipped with an X-ray source with maximum voltage and current of 160 kV and 5 mA respectively. The focal spot size was 1 mm. The detector system used was Linear Array Detector (LAD) with resolution of 0.4 mm/pixel.

The voltage applied in this work was at 60 kV and a current of about 2.0 mA. The source-detector and source-sample distance were 50 cm and 50 cm respectively. A fan X-ray beam scanned the sample with rotation increments of $0.5°$ s^{-1} for each step. The procedure was repeated until $360°$ of rotation and a total of 720 projections were obtained to be used in the image reconstruction. The software used for the image reconstruction was OCTOPUS and the integration time was 1 s.

The AE system used consists of AE four channel systems as shown in Figure 11.3. The system comprises of AE mechanical pencil with its pencil lead, sensor R15, pre-amplifier 2/4/6, power supply, couplant and software for data replay and analysis.

FIGURE 11.2   X-ray computed tomography system.

FIGURE 11.3   Acoustic emission system.

For each material, during the impact tests, every impact event was recorded with the acoustic emission system. The AE signals were detected using R15D differential sensor with a resonant frequency at 150 kHz and diameter dimension of 18 mm. The bandwidth of 12 kHZ–382 KhZ were used. For every test, the output from the sensor was amplified by 40 dB. PAC 2/4/6 preamplifiers were used throughout this study. The sensor was mounted to the surface of the hybrid composites before it was clamped at the drop test rig as shown in Figure 11.4. The analysis on frequency was focused throughout this work.

In the UT method, a probe with frequency 1 MHz was used as shown in Figure 11.5. Gel was used as the couplant as the medium for the signal transmission. Calibration was conducted to obtain the time travel and velocity of the material.

All the scanning results were stored in the computer. The scanning was performed from the top to the bottom of the samples. Different colours appeared on the image

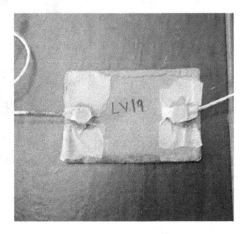

FIGURE 11.4   AE sensors attach to the composites.

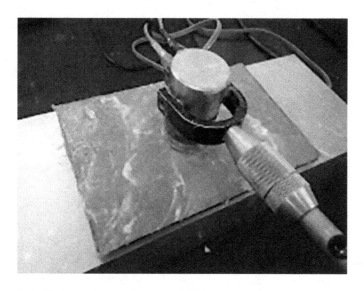

FIGURE 11.5 Single probe with frequency 1 MHz.

representation of the thickness difference. Blue colour was calibrated as thickness sample of 20 mm while green and yellow represented the thin area.

11.3 RESULTS AND DISCUSSIONS

11.3.1 LOW-VELOCITY IMPACT RESULTS

From the low-velocity impact testing results, Figure 11.6a–d show the force displacement curve for the impacted samples at four different impact energy levels.

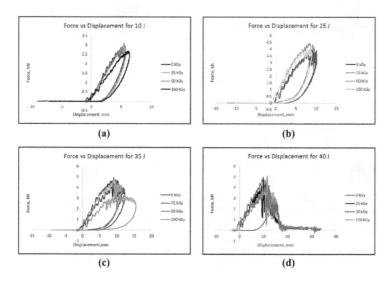

FIGURE 11.6 Force-displacement curves at different impact energy levels.

The force-displacement curve shows the damage progression on the tested samples. The closed curve in Figure 11.6 a–c show that no penetration occurred after the impact testing, for samples without and with radiation. However, at 40 J impact energy, penetration occurred at radiated samples, but irradiated samples did not experience any penetration. This can be seen from Figure 11.6d, where the curve for radiated samples is a close curve, while the curve for irradiated curve is an open curve.

Since most of the samples did not experience any penetration during the impact test, it is hard to determine the level of damage experience by the samples. Therefore, to evaluate the internal damage on these samples, CT, UT and AE methods were chosen. AE was conducted in-situ during the impact event while CT and UT were conducted after the impact event.

11.3.2 CT RESULTS

Figures 11.7–11.10 display the CT images for different impact energy for specimens with and without radiation. From the CT image, the internal damage could be observed, such as voids, delamination and fiber breakage. Voids usually occurred due to fabrication process while delamination and fiber breakage occurred due to the

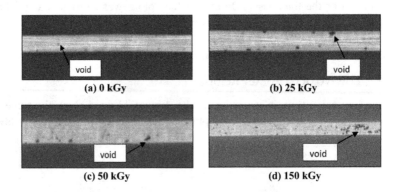

FIGURE 11.7 a–d CT images at different radiation dose with 10 J impact energy.

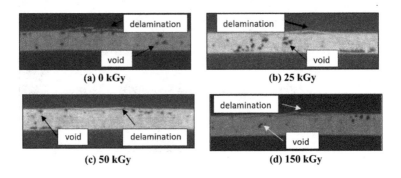

FIGURE 11.8 a–d CT images at different radiation dose with 25 J impact energy.

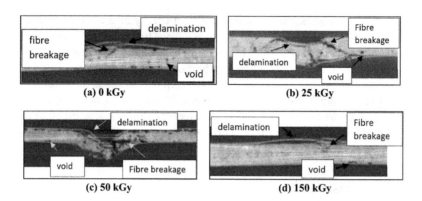

FIGURE 11.9 a–d   CT images with various radiation dose with impacted energy of 35 J.

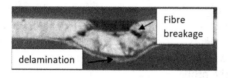

FIGURE 11.10   CT image of without radiation dose with 40 J impacted energy.

impact event. The initiation of damage modes were influenced by the nose and shape of the impact, face sheet thickness to span ration and panel support condition [17].

Figure 11.7a–d show CT images of specimens when impacted with 10 J of impact energy. Figure 11.7a is the specimen without radiation, Figure 11.7b is the specimen with radiation dose of 25 kGy, Figure 11.7(c) the specimen with radiation dose of 50 kGy and Figure 11.7d the specimen with radiation dose of 150 kGy. From Figure 11.7, only voids were detected in the internal structure of the composites. The voids in the internal structure are believed to have occurred during the fabrication process and not from the impact. There was no other impact damage occurred in the specimens such as delamination and fiber breakage. It can be justified that the hybrid combination is able to withstand the impacted energy at 10 J for specimens without and with radiation dose.

At 25 J impacted energy, delamination was detected in the internal structure of the specimens regardless whether the specimens were irradiated or not irradiated. From Figure 11.8a–d, only delamination was observed.

Figure 11.9a–d show CT images of specimens when impacted with energy of 35 J on specimens that were radiated and not radiated with gamma radiation. At energy of 35 J, more damages were detected compared to the lower impact energy level. Delamination and fibre breakage were observed in the specimens that were irradiated and not irradiated. However, the damages, such as fibre breakage and delamination, in specimens that were not irradiated were not severe as the fibre breakage and delamination that occurred in irradiated specimens. From Figure 11.9(c), it is observed that more fibre breakage occured. This may be due to many voids in the specimens, hence deeper dents were observed from the image. For that reason, the peak in the force displacement curve for impact energy 35 J at 50 kGy is lower as compared to impact force at 35 J for other specimens that were irradiated with different radiation doses and specimens that were not irradiated as shown in Figure 11.6(c).

From Figure 11.10, it is noted that fibre breakage and delamination existed, and the impact resistance at 40 J showed a close loop which indicated that the hybrid composites without radiation are stronger and able to withstand the impact energy at 40 J.

From the results of the CT images, at different impact energy, different damage modes occurred. This shows that the different damage mode took place due to the factor of constituent properties and loading condition [18].

11.3.3 Ultrasonic C-scan Results

The image from ultrasonic C-scan images represented defects at the impacted surface and a few millimeters depth from the surface. The ultrasonic C-scan testing was done on specimens that were irradiated and not irradiated and impacted only up to 35 J of impact energy. The scan was not conducted at samples impacted at 40 J impact energy because the sample was deeply indented, although it was not fully penetrated yet. The deeply indented surfaces affected the ultrasonic signal. This is because the probe was unable to detect the signal due to no contact between the dented surface and the probe.

However, the images obtained from C-scan were limited. Although ultrasonic C- scan is widely used for damage evaluation in composite materials, there were still setbacks in using this method for this material. Composite materials are heterogeneous materials, hence, the ultrasonic waves suffer from loss of signal due to large attenuation and scattering [19], thus affected the damage interpretation.

Figure 11.11a–d show the images from the C-scan of samples impacted at energy level of 10 J. The images showed that there are damages on the impacted samples. The blue colour represents the background and there were no flaws in the blue colour area. However, the yellowish colour area represents flaws that exist at the area due to impacted surface as shown in Figure 11.11a and 11.11b. Figure 11.11c displays a black colour image at the centre of the specimen. The black colour indicates that there was signal loss at that area. The signal loss is due to existence of air gap at the specific area or no contact between the probe and the surface. Hence, the black area appeared in the image because ultrasonic waves are not able to pass through air. The existence of an air gap and no contact may be due to delamination or a dent at the surface and subsurface of the impacted area.

At impact energy 25 J, all hybrid composites that were irradiated and not irradiated displayed images with a black colour area as shown in Figure 11.12a–d. This indicates that there exists a hole or dent at the surface of the hybrid composites. Existence of a dent or hole at the surface prevents the wave passing through the hybrid composites.

Figure 11.13 a–d)show ultrasonic images on an impacted surface at energy 35 J. From Figure 11.13, the black area obviously appeared in all images. The black area indicates that dents appeared on the surface of the hybrid composites. In principle, as the impacted energy increases the damage size will also increase. This trend could be observed at hybrid composites without radiation and hybrid composites irradiated at

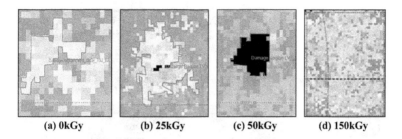

(a) 0kGy   (b) 25kGy   (c) 50kGy   (d) 150kGy

FIGURE 11.11 a–d Ultrasonic image on impacted surface at 10J.

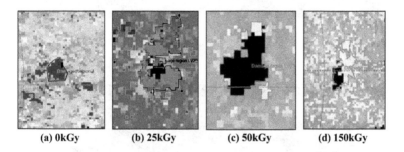

(a) 0kGy   (b) 25kGy   (c) 50kGy   (d) 150kGy

FIGURE 11.12 a–d Ultrasonic images on impacted specimens at energy 25 J.

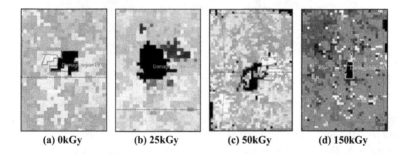

(a) 0kGy   (b) 25kGy   (c) 50kGy   (d) 150kGy

FIGURE 11.13 a–d Ultrasonic images on impacted surface at 35 J.

25 kGy radiation dose. However, at radiation 50 kGy and 150 kGy the hybrid composites did not obey the stated principle. This may be due to the material characteristics of Kevlar and oil palm empty fruit bunch, which is an absorbent material [20, 21]. In the absorbent material, the signal is being absorbed in the material and not being reflected. Hence, the probe was not able to detect signals reflected from the material.

11.3.4 AE RESULTS

In AE method, the signal was captured simultaneously with the impact. The signal observed is the transient AE signal where it can be a burst type signal or pulse as shown in Figure 11.14. This signal agrees with the impact phenomena where impact

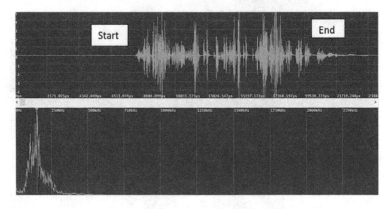

FIGURE 11.14   AE signals recorded from the AE system.

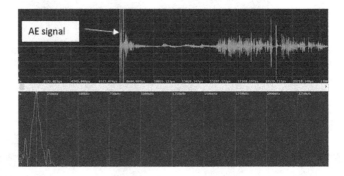

FIGURE 11.15   AE signal from impacted energy 10 J.

damage caused an abrupt and permanent change in the materials. The pulse is identified by the start and end of the signal.

Analysis of the recorded AE signals were explored, which provides more information on the events and mechanisms that lead to failure. In analysing the signal, fast
Fourier transform (FFT) imbedded in the system were calculated from the captured
AE signal. The frequency range obtained after the treatment with FFT provides information on the damage mechanism that occurred in the hybrid composites.

At energy 10 J, the frequency obtained during the impact was 167 kHz after the
FFT treatment as shown in Figure 11.15. This frequency shows that only matrix
cracking occurred in the sample [22] which is in range 110–260 kHz. The results
depicted that the earliest damage mechanism detected using the AE system is matrix
cracking. The result shows that in this hybrid, matrix cracking was the first damage
experienced by the material due to the frequency detected by the system. These
results agree with the results found in [22] where the frequency range for matrix
cracking is found to be in range 110–260 kHz. Comparing the AE results with the CT
images, for 10 J of impacted energy, there was no delamination found in the CT
images and this agreed with the pulse obtained from the AE system. However, with

the AE system matrix, the AE system is capable to also detect matrix cracking, where this matrix cracking cannot be detected by using the CT testing.

At energy 25 J, matrix cracking damage was detected as shown in Figure 11.16. The frequency for the matrix cracking during the impact detected was at 152 kHz. This range falls in the matrix cracking damage [22]. With further scanning, frequency at 235 kHz was detected. From the analysis, it was observed that the signal found was like the high spike signal during the matrix cracking mechanism. According to Mahdian et al. [22], frequency in range 230–400 kHz shows that debonding or delamination occurred. It can be concluded that at 25 J, delamination has started to occur. These results were similar to the results found in CT images at 25 J where, from the CT images, it shows that debonding or delamination occurred. The frequency 235 kHz was detected after detecting the frequency of 152 kHz for matrix cracking. This directly shows the damage progression on the impacted samples, where at the beginning, matrix cracking occur followed by the delamination.

Figure 11.17 shows the AE signal from impacted energy of 35 J, where similar damage mechanism as in impact energy 25 J was observed. From the scanning, matrix cracking was detected at 121 kHz. Further scanning found at the frequency of

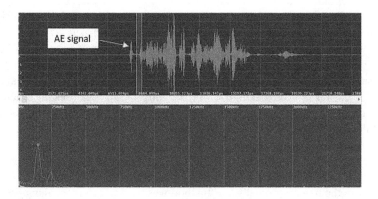

FIGURE 11.16   AE signal from impacted energy 25 J.

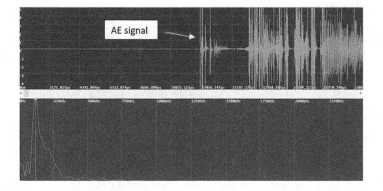

FIGURE 11.17   AE signal form impacted energy 35 J.

209 kHz that delamination had occurred. [21]. During the debonding mechanism there were lower spike signals; however, the frequency of the signal was higher, in the range of 230–400 kHz. However, comparing the AE results with the CT images at impact energy 35 J, CT images shows that the samples experienced fibre breakage. According to Mahdian et al. [22], the fibre breakage mechanism only occurred at frequency range 340–500 kHz. Unfortunately, in this work, the ultrasonic C-scan machines can only detect frequencies up to range of 260 kHz–270 kHz. This may be due to the limitation of the sensor used during the AE testing. A wide band sensor is suggested to be used in this work in order to obtain higher frequency range.

11.4 CONCLUSION

In this research, it was found that the impact damage differs for every impact energy level. The damage becomes more severe as the impact energy level increases. In the laminated hybrid composites, three major defects were observed. Those damages were matrix cracking, delamination or debonding and fibre breakage. The appearance of the damage depended on the impact energy level. In this research, the samples were radiated with different gamma radiation doses. However, from the results, it can be concluded that there was no significant improvement in terms of impact resistance of the composite even after the samples being radiated.

In evaluating the impact damage, various methods were applied, namely computed tomography (CT), ultrasonic C-scan (UT) and acoustic emission (AE). From the results, it was found that CT method was able to detect delamination and fibre breakage on the impacted samples. For the ultrasonic C-scan testing, there were drawbacks in using it to detect damages on the samples. The probe was not able to detect the signal on heavily dented samples due to loss of signal. Therefore, it can be concluded that ultrasonic C-scan testing can only be applied on impacted samples that are not heavily dented/damaged. Other than that, according to the literature it was found that oil palm EFB is a good absorber for sound waves, hence loss of signal occurred in the material. Finally, AE method was used to detect the damages in this research. From the results, it was found that this method managed to detect matrix cracking and delamination on the impacted samples. However, the AE method did not manage to detect fibre breakage due to the limitation of the sensor.

From the NDT results obtained, there was no single NDT method that was capable of detecting all the internal damages. This is due to the limitation of the methods and characteristics of the materials. Hence, it can be concluded that the internal impact damages on the impacted samples can be verified using this combination of NDT methods which are CT, ultrasonic C-scan and AE.

ACKNOWLEDGEMENT

This work is supported by UPM under Geran Putra Berimpak GPB 9668200. The authors would like to thank the Department of Aerospace Engineering, Faculty of Engineering, Universiti PutraMalaysia and Laboratory of Bio-composite Technology, Institute of Tropical Forestry and Forest Products (INTROP), Universiti Putra Malaysia (HICOE) for the close collaboration in this research.

REFERENCES

1. Taraghi, I, Fereidoon, A, Taheri-Behrooz F. "Low-velocity impact Response woven Kevlar/epoxy laminated composites reinforced with multi-walled carbon nanotubes at ambient and low temperatures." *Materials and Design* (2014); 53: 152–158.
2. Aktas M, Karakuzu R, Arman Y. "Compression-after impact behaviour of laminated composite plates subjected to low velocity impact in high temperatures". *Composite Structures* (2009); 89(1): 77–82.
3. Richardson M, Wisheart M. "Review of low-velocity impact properties of composite materials." *Composites Part A: Applied Science and Manufacturing* (1996); 27A: 1123–1131.
4. Sivakumar D, Kathiravan S, Selamat MZ et al. "A study on impact behaviour of a novel oil palm fibre reinforced metal laminate system." *ARPN Journal of Engineering and Applied Sciences*, (2016); 11:2483–2488.
5. Zainudin ES, Yan LH, Haniffah WH, Jawaid M, Alothman OY. "Effect of coir fiber loading on mechanical and morphological properties of ol palm fibers reinforced polypropylene composites." *Polymer Composites* (2014);35(7):1418–1425.
6. Channabasavaraju S, Shivanand H, Kumar SS. "Investigation of low velocity impact properties of Kevlar fiber reinforced polymer matrix composites." *International Journal of Engineering Research & Technology* (2013);2(10):680–682.
7. Ying S, Mengyun T, Zhijun R et al. "An experiemental investigation on the low-velocity impact response of carbon-aramid/epoxy hybrid composite laminates." *Journal of Reinforced Plastics and Composites* (2016);36(6):1–13.
8. Gustin J, Joneson A, Mahinfalah M, Stone J. "Low velocity impact of combination Kevlar/carbon fiber sandwich composites." *Composite Structures* (2005); 69(4): 396–406.
9. Shaari N, Jumahat A, Razif M. "Impact resistance properties of Kevlar/glass fibrt hybrid composite laminates." *Jurnal Teknologi* (2015);76(3):93–99.
10. Imielinska K, Guillaumat L. "The effect of water immersion ageing on low-velocity impact behaviour of woven aramid-glass fibre/epoxy composites." *Composites Science and Technology* (2004);64(13–14):2271–2278.
11. Ismail MF, Sultan MTH, Hamdan A, Md Shah AU. "A study on the low velocity impact response of hybrid kenaf-Kevlar composite laminates through drop test rig technique." *Bioresources* (2018);13(2):3045–3060.
12. Fidan S, Snmazcelik T, Avcu E. "Internal damage investigation of the impacted glass/aramid fiber reinforced composites by micro-computerized tomography". *NDT&E International* (2012);51:1–7.
13. Woo S, Kim T. "High strain-rate failure in carbon/Kevlar hybrid woven composites via a novel SHPB-AE coupled test". *Composites Part B: Engineering* (2016); 97: 317–328.
14. Jaroslav Z, Michal S, Petr D, Martin B. "Acoustic emission during tensile testing of composite materials reinforced carbon and aramid fibers". *Mendel Net* (2015), 568–572.
15. Maleki H, Akbar AF. "On the Nondestructive Evaluation of Composite Structure by Liquid-Crystal Thermography". In *17th World Conference on Nondestructive Testing*, Shanghai, China. (2008). p. 25–28.
16. Destic F, Petijean Y, Massenot S, Mollier J, Barbieri S. "THz QCL-based active imaging dedicated to non-destructive testing of composite materials used in asronautics". *InTerahertz Emitters, Receivers, and Applications. 2010. International Society for Optics and Photonics Proceedings of SPIE* (2010).
17. Razali N, Sultan MTH. "The study of damage area and non-destructive testing on glass fibre reinforce polymer after low velocity impact event". *Applied Mechanics and Materials* (2015); 874–880.

18. Patil S, Reddy DM, Reddy M. "Low velocity impact analysis on compositee structures - A review". In *AIP Conference Proceedings*, (2018);1943:2–9.

19. Tsao C, Hocheng H. "Computerized tomography and C-Scan for measuring delamination in the drilling of composite materials using various drills". *International Journal of Machine Tools & Manufacture* (2005);45(11):1282–1287.

20. Samsudin EM, Ismail LH, Kadir AA, Mokdar SS. "Comparison on acoustic performance between dust and coir form empty fruit bunches (EFB) as sound absorption material". *Jurnal Teknologi* (2016);78(5):191–196.

21. Hee OK, Putra A, Nor MJ, Selamat MZ, Ying LZ. "Sound absorption performance of oil palm empty fruit bunch fibers." In *23rd International Congress on Sound and Vibration, ICSV* (2016).

22. Mahdian A, Yousefi J, Nazmdar M, Zarif Karimi N, Ahmadi M, Minak G. "Damage evaluation of laminated composites under low-velocity impact tests using acoustic emission method". *Journal of Composite Materials* (2017);51(4):479–490.

12 Mechanical Behaviour of Sandwich Composites in Automotive Applications

Amol Bhanage, Saurabh Bait and Ramesh Sakhare

Marathwada Mitra Mandal's Institute of Technology, Pune, India

CONTENTS

12.1 INTRODUCTION

Advanced composite materials have been used in the transportation industry for a long time (C. McCarthy, 2015). Sandwich composites, which are a special form of composites, are finding an increasing role in the aerospace structure, ship structure and high-speed submarines, buses, rail vehicles, trucks and cars. The main reason behind this is that sandwich composite structures have a high strength-to-weight ratio, high energy absorption, superior stiffness at minimum weight, impact resistance, and are lightweight. In order to get all the benefits from sandwich composites in automotive applications, the possible areas of use are in vehicle floors, wheels, bumpers, roof structures, engine hoods, door panels, etc. This resulting weight reduction is more useful in electric vehicles.

DOI: 10.1201/9781003200994-12

The sandwich composite structure consists of a core of low strength and two thin skin or face sheets of high strength. The thin face sheets or skins are made from metal alloys, or polymer sheet or fibre reinforced composites. The core is thick, made of materials such as polymer or metal foam, honeycomb, balsa wood, web core, etc. This core and face sheets are bonded together to facilitate the load transfer mechanism. The most common core materials are thermoset polymer foams to accomplish high temperature tolerances, although thermoplastic foams are used in all applications. The most common are polyurethane (PUR), polyvinyl chloride (PVC), polymethacryl-imide (PMI), polyetherimide (PEI), polyphenols (PF) and polystyrenes (PS). From these, PMI and PEI are high performance cores used in aerospace applications. Honeycomb structures are known for low weight and high bending stiffness and are widely used under tensile and bending loads. These structures are used in place of conventional materials under high loading. Honeycomb cores clearly dominate over alternative materials. These honeycomb cores are based on aluminum alloys, i.e., Al 3003, Al 2024, Al 5052, etc., and phenolic resin bathed aramid fibre paper, the latter under the trade name Nomex (K. F. Karlsson and Tomas Astrom 1997).

Face sheets are usually made of metal or a fibre-based composite. In sandwich composite structure, face sheets play a role to withstand the bending movement on the panel or to resist the tensile and compressive loads created in the opposite skin when the panel is subjected to bending forces. As mentioned, face sheets or skins are normally made of fibre-based composite such as glass fibre reinforced composite (GFRP) or carbon fibre reinforced composite (CFRP) materials. GFRP or CFRP have high flexural strength, flexural stiffness and high specific strength. The commonly used core materials are open-cell structured foam, closed-cell structured foam, composite honeycomb and balsa wood. From these, balsa wood and foam cores are soft and will crush under high compressive loads, while honeycomb cores have high compressive strength.

12.2   AN AUTOMOTIVE PERSPECTIVE

The advantage of foam, honeycomb and balsa wood sandwich structures are used in automotive to reduce weight under physical loads such as bending, twisting or buckling and torsion, etc. In preference of sheet metals, these structures reduced the amount of emission and power needs. Vibrational and acoustic studies show that the sandwich panel can reduce the weight approximately 60–70 % by keeping same damping performance (Deniz Hara and Ozgen 2016).

12.2.1   AUTOMOTIVE BUMPER

Crashworthiness is an important design parameter for automotive applications. The use of sandwich structure in the front bumper reduced its weight by 18%. The sandwich bumper absorbs more energy and sustained deformation for a long time. The use of sandwich composite panel in automobile front bumpers reduced the risk of injuries to the occupant by 40 percent (Amarnath Donga, et al. 2011).

12.2.2 BODY PANEL

Sandwich composite construction is used in making the bodies of electric vehicles, bus, trucks, train and in such applications that can take advantage of its lightweight yet structurally robust feature. In modern trucks, sandwich composites are used in doors, ceilings, side walls and floor panels components. The advanced synthetic fibres, PVC or foam can also be used in other applications. In electric vehicles, the frame body is made of polypropylene honeycomb core and hybrid carbon-glass fibre composite skin construction. This is best substitute in automotive as compared to GFRP and CFRP fibre reinforced composites (S. Cahyono 2016). In bus applications, thermoplastic composite body panel is best alternative to conventional aluminium skin with supporting steel bars. Thermoplastic sandwich structures used in the bus body panel with polypropylene honeycomb core and E-glass fibre/polypropylene sheets have high weight savings (around 55%), high strength and high energy absorption benefits (Haibin Ning et al. 2007).

12.3 QUASI-STATIC BEHAVIOUR

The sandwich composites are manufactured differently from skin, and core materials with their different composite structures are studied here for their probable application in automotive structure or components. In quasi-static behaviour study, mostly experimental studies have been done on three-point bend, four-point bend and compression tests. The three-point sandwich structure bending test is performed on aluminium skin sheet, foam core, skin/sheet-foam-skin/sheet plies without bonding and sandwich composites in order to understand behaviour and its advantages. The static strength of plies of sheet-foam-sheet is very nearly the same as the strength of foam core. This predicted the importance of the bond between the core and the face sheet; the combined response of a sheet and a skin improves the strength of the structure (F. P. Yang et al. 2015). The sandwich composites are made from pure aluminium face sheets ($AlMg_3$) and open cell (Nomex) aramid honeycomb core. The structures are studies for different densities of 48, 80, 128, and 144 kg/m^3 under four-point flexural test. These structures showed maximum ultimate loading for high density cores and showed ductile behavior compared to sandwich structures with low core densities. L cell direction can carry higher maximum loads and stiffness than W cell direction (K. Boualem 2013) (Table 12.1).

Sandwich samples were made of carbon woven/epoxy skin (CFRP) and Nomex core made of hexagonal cell. In static behaviour, as shown in Table 12.1, Phase I was observed to be related to the compression and tensile behavior of the skin, with the load-displacement relationship being linear up to the maximum reached. The impetuous load that decreased in Phase II is related to the precipitous failure of the top sandwich skin under compression. Phase III depends on the nonlinear behaviour of the honeycomb core (W. Boukharouba, et al. 2014) (Figure 12.1).

Sandwich structure made from glass fibre /epoxy skin (GFRP) skin and Nomex honeycomb core are also studied for different temperature failure behaviour under three-point bend tests (Eynali and Roknizadeh 2019). The mechanical properties of sandwich composites made from honeycomb polypropylene core and glass fibres/

TABLE 12.1

Quasi-static Properties for W and L Direction Nomex Sandwich

Sr. No.	Types of Nomex honeycomb sandwich	Density kg/m³	Load N	Deflection mm	Reference
1	L direction	144	10,000	82	K. Boualem
2		128	9,500	70	2013
3		80	8,000	22	
4		48	6,500	20	
5	W direction	144	9,000	40	
6		128	7,800	32	
7		80	6,000	18	
8		48	3,800	15	

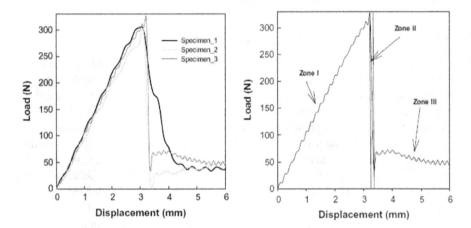

FIGURE 12.1   Load–deflection Curves of CFRP Skin/ Nomex Core.

polyester resin face sheet depends on variation in the core thickness and intermediate layer. Its properties are increased (as shown in Table 12.2 and 12.3) with increasing thickness of the core and intermediate layers (J. Arbaoui, et al. 2014).

Sandwich composites are also made from fibre reinforced composites (FRP) face sheet and honeycomb core is usually made of aluminum. In composite structure, GFRP is face sheet and aluminium 5052- H32 core, consisting of five phases of static failure (Figure 12.2). In Phase I and II, linear elastic behaviour of the composite panel is observed until the ultimate load. Phase III showed an abrupt decrease in load of the honeycomb. The load was slightly decreasing later, corresponding to the structural stabilisation in Phase 4 and, finally, in Phase 5, the load carrying capacity reduced due to bending of cell walls, laminar shear failure of facing, and core shear failure (Muzamil Hussain, et al. 2019).

Polymeric foams such as PVC, PET or PMI foam have been used as cores for automotive sandwich composites structures. Using polymeric foam, the sandwich structure becomes lightweight, leads to increases in its energy absorption and

TABLE 12.2

Mechanical Properties Honeycomb Polypropylene Core with Different Thickness

Sr. No.	Core Thickness mm	Load N	Facing stress MPa	Core Shear Stress MPa	Bending Stiffness (10^5) N/mm^2	Shear Stiffness N	Reference
1	10	350	68.20	0.45	213.75	3388	J. Arbaoui,
2	20	480	49	0.33	527	6174	et al. 2014
3	40	594	31	0.21	855	11767	

TABLE 12.3

Mechanical Properties of Honeycomb Polypropylene Core with Different Core Layer

Sr. No.	Core Layer	Load N	Facing stress MPa	Core Shear Stress MPa	Bending Stiffness (10^5) N/mm^2	Reference
1	Single Core	740.67	38.71	0.25	855	J. Arbaoui, et al.
2	Double Core	792.68	41.43	0.28	1058	2014
3	Triple Core	903	47.20	0.31	1054	
4	Quadruple Core	1022	51.74	0.33	1218	

improves the crash properties of the structure. Additionally, its good thermal insulation properties are especially beneficial for electric cars as no engine is used to heat up the passenger department. Sandwich composites are used with different face sheet/skin and core configurations as shown in Table 12.4. The core material is PVC foam and glass/polyester resin (CSM) is used as a face sheet. Experimental three-point bending tests of this configuration predicted that the mechanical properties would be better with decreasing core thickness and number of layers (Table 12.5) (Gonabadi, et al. 2016).

Structures were composed of a PVC foam core and polymer reinforced fibreglass (GFRP) with configurations (80:80:80), (60:100:60), (100,60:100) and (60,80:100). A quasi-static study of flexural tests, as shown in Figure 12.3, showed that low density core configuration reduced the failure probability of the sandwich structure in comparison to uniform and homogenous layered core configuration (Cihan Kaboglu, et al. 2018).

PVC foam core with glass fibre/epoxy skin composites were tested for their load carrying capacity. The increases in either a number of plies (skin layers) or core density increase its bending resistance (C. Uzay, et al. 2017). The bending stiffness of the same sandwich composites having cross-ply laminate configuration has more bending stiffness than unidirectional laminate configuration. The effect of core thickness

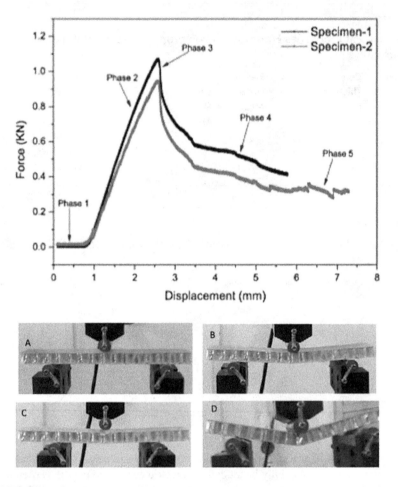

FIGURE 12.2 Quasi-static behavior of FRP sandwich composite under three-point bend.

TABLE 12.4
Configuration of Polymeric Foams Sandwich Construction (SC)

Sr. No.	Sandwich Construction	Layers (No.)	Facing Sheet construction (Thk, mm)	Core Construction (Thk, mm)	Reference
1	SC 1	5 layers	2 (2.25 Thk)	3 (1.15 Thk)	J. Arbaoui, et al.
2	SC 2	3 layers	2 (2.00 Thk)	1 (4.00 Thk)	2014
3	SC 3	3 layers	2 (2.00 Thk)	1 (0.75 Thk)	

on bending behaviour of polyurethane (PU) foam and E-glass fibre–reinforced epoxy also indicated that an increase in core thickness leads to a decrease in axial stress on the skin and shear stresses in the core. The higher thickness also increased flexural strength and rigidity without comprising a weight (Yash Gupta, et al. 2020).

TABLE 12.5

Experimental Properties of Sandwich Composites

Sr. No.	Sandwich Construction	Flexural Strength (MPa)	Flexural deflection (mm)	Reference
1	SC 1	118	5	J. Arbaoui, et al.
2	SC 2	90	4.4	2014
3	SC 3	197	10.9	

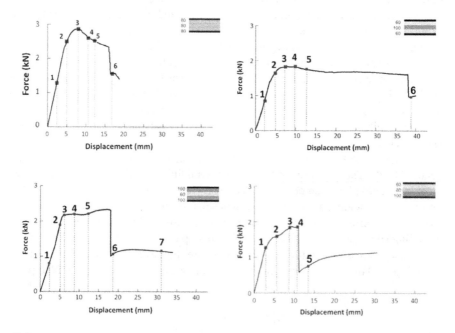

FIGURE 12.3   Effect of core densities on quasi-static flexural behaviour.

Thermoplastic foam, i.e., PET foam, sandwich structure offers significant advantages for the automotive industry (T. Neumeyer, et al. 2017). The static behavior of the PMI foam core sandwich structure was also studied. Polymeric foam PMI with fibre-reinforced skin composite sandwich was studied with unidirectional, quasi-isotropic, cross-ply and angle-ply, etc. face sheet configuration numerically and experimentally. It was found that in quasi-static bending, the unidirectional orientation has a high flexural stiffness along with fibre orientation, while the angle-ply structure has a lower flexural stiffness (Fa Zhang, et al. 2013).

The application of closed-cell polyurethane foam (PU) foam core and acrylic sheet skins sandwich composites can be found in the low load carrying automotive components. The compression test of these samples has maximum strength up to 2 MPa (M. Ansari et al. 2015). Unidirectional carbon/glass fibre as face sheets and aluminium 5052 as hexagonal honeycomb core sandwich composite structure were studied using FEA software. Von Mises stresses and deformation are calculated from finite element analysis predictions and were found to decrease with an increase in the face sheet thickness (Shubham Upreti, et al. 2020).

12.4   DAMAGE BEHAVIOUR UNDER STATIC LOADS

Sandwich composites made of open-cell aluminium foam core and Al-2024 aluminium skins failed under quasi-static loading several modes of fracture i.e., shear failure of core at central and boundary, upper face wrinkling, bottom face fracture, or interfacial failure (Lin et al. 2010). The ruptures of the composite sandwich made of PVC foam and fibreglass/ epoxy resin skins depend upon the type of face sheet/ skin (unidirectional or cross-laminate). In the case of unidirectional laminates, the fracture occurs (Figure 12.4) due to the delamination between skin and core, foam shear and finally lower skin foam cohesion at the bottom. For cross-ply laminates, a fracture occurred in the upper part of the skin near the central load (A. Chemami et al. 2012). Tests and numerical studies have shown that the polymer foam with a fibre-reinforced skin sandwich composite failed with catastrophic damage due to core shear failure and skin compression, followed by debonding between the layers, skin, and core.

The composite structure of honeycomb core consists of five major failures under bending loads. These failure modes are face wrinkling, core shear under shear stress, local indentation between core and face, face yielding and intra-cell dimpling. The observed face yielding at maximum normal stress reached the elastic limit. Similarly, wrinkling occurred when the maximum normal stress of the skin reached critical wrinkling stress of the skin. The shear failure depicted in Figure 12.4 occurs in the core if shear stress reached shear strength of core. Figure 12.5 showed that the principal degraded modes of Nomex honeycomb sandwich structures under a three-point

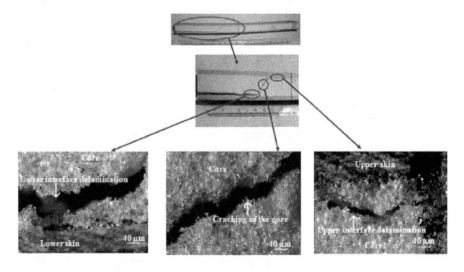

FIGURE 12.4   Fracture surface of unidirectional composites.

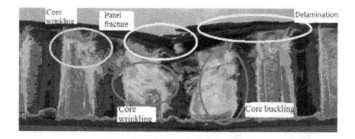

FIGURE 12.5   Failure modes of Nomex honeycomb sandwich composites.

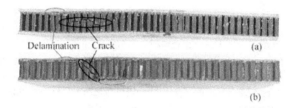

FIGURE 12.6   Failure modes of Nomex honeycomb in (a) W direction specimen (b) L direction specimen.

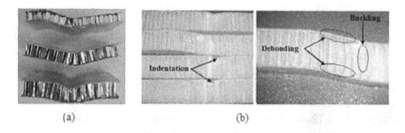

FIGURE 12.7   Bending failure modes of (a) 10 mm, 15mm and 20 mm thick specimen (b) honeycomb polypropylene core and glass fibre/polyester resin face sheet.

static bending load are core wrinkles, core buckling, delamination and face-sheet fracture etc. (X. Wu, et al. 2019). Similarly, Figure 12.6 showed that Nomex honeycomb sandwich specimens rupture in the L and W directions due to core shear failure and delamination (core/face sheet), respectively.

In three-point bending testing, the sandwich structures lead to local indentation, buckling and debonding of skins and core at the centre. Figure 12.7a shows specimens with 10 mm, 15 mm and 20 mm thickness have complete bending, partial local buckling and wrinkling damage behaviour, respectively. Figure 12.7b shows bending failure of a honeycomb polypropylene core and glass fibre/polyester resin face sheet.

The failure mechanism (Figure 12.8) of the low density GFRP /polymeric foam sandwich structure is that core crushing, shear cracking, and skin/core debonding occurs easily. This type of sandwich provides an easy fracture path to grow out and cause debonding (Cihan Kaboglu, et al. 2018).

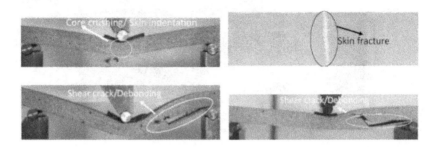

FIGURE 12.8   Fatigue failure mechanism of low density GFRP /polymeric foam sandwich structure.

Face sheet damage under quasi-static loading is matrix cracking, matrix delamination, fibre breakage, etc. These established failure modes are dependent on the face sheet configuration. In unidirectional face sheet sandwich foam composites, a small debonding was observed in transit of bottom core/face interface, whereas in crossply they have a larger debonding size along the bottom core/face interface. The orientation of fibre plays an essential role in damage behaviour of face/core interface.

12.5   FATIGUE BEHAVIOUR

The fatigue strength of sandwich structures mainly depends on the shear strength of the structure. Sandwich structures are primarily tested with a three-point bending load. In three- point bending loading tensile, compressive and shear stresses are noticed in the core structure. During the fatigue test, the sandwich samples are repeatedly subjected to tensile and compressive stress in top and bottom face sheets, respectively. This stress accumulates shear stresses in the sandwich construction. The ability of the core to compensate for these stresses determines the fatigue life of sandwich composites. The failure under fatigue stress is complex and associated with the behavior of multiple cracks due to the anisotropic behavior of sandwich composites. Fatigue failure of the foam core sandwich composites is controlled by core failure. The crack propagates in three stages: core-face sheet debond, then core shearing, followed by further core-face sheet debonding (Figure 12.9) (Nitin Kulkarni et al. 2003).

The progression of damage under static and fatigue loading is identical, but with the exclusion that the fatigue loading at a certain level of stress will cause additional damage, dependent on the frequency of fatigue cycle. Polyurethane (PU) sandwich constructions are popular for their stiff and lightweight constructions. The fatigue behaviour of polyester/glass-PUF-polyester/glass, epoxy/glass-PUF-polyester/glass and epoxy/glass-PUF-epoxy/glass for frequencies of 1, 3 and 5 Hz is studied. The epoxy/glass-PUF-epoxy/glass sandwich structure has good bending strength and can prevent the skin from bending on the compression side and also able to bear the high compressive stresses encountered during fatigue loading. These structures also adapt to higher fatigue stresses (S.C. Sharma et al. 2004). The fatigue of woven roving, chopped strand mat (CSM), stitch bond mat (SBM) E-glass fabric with vinyl ester resin polyurethane foam sandwich composites for different frequencies predicted

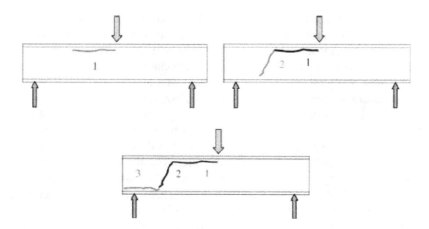

FIGURE 12.9 Fatigue damage event in sandwich composites.

that the composites did not fail at low frequencies, but they failed at higher frequencies due to delamination and core cracking. The woven roving sandwich composites showed the highest fatigue strength. Fatigue crack failure is different for composite materials with different densities; it is located closed to the interface between core/face sheet at low density and in the center of the core for high density sandwich composite structures (Table 12.6) (Manujesh and Rao 2013).

The sandwich structure is also made with PMI foam core and EN-AW 5754 aluminium skin sheets. The degree of fatigue fractures of this sandwich structure is influenced by the thickness of skin sheet, the thickness of foam and span length. The thin skin sheet sandwich structure is failed by local depression, after which the core failed due to bending and shear loads. The fatigue load under force control mode was tested with 0.9, 0.85, 0.8, 0.75 and 0.65 load levels (F. P. Yang et al. 2015).

The fatigue test frequency and fluctuating loads play an important role in determining fatigue strength of the PU foam sandwich structure. The fatigue strength

TABLE 12.6
Failure Modes Observed PU foam Sandwich Composites

Sr. No.	Sandwich Type	Density of PU Foam (kg/m³)	Cycles to failure		Failure mode observed	Reference
			3 Hz	9 Hz		
1	Woven	100	16350	8186	Debonding,	Manujesh and
2	Roving	200	21360	8900	Core crack,	Rao 2013
3		300	27980	14050	Delamination	
4	CSM	100	14648	1923		
5		200	16650	3219		
6		300	17507	4885		
7	SBM	100	21354	22150		
8		200	30050	26064		
9		300	16156	3455		

increased with increasing density of the PU foam and degradation of stiffness increased with increasing test frequency. The fatigue behaviour of PVC foam core and chopped strand mat (CSM) glass/polyester resin under three-point bend are comparatively the same as observed in quasi-static loading. The fatigue performance of the PU foam and glass/epoxy composite face sheet under impact load are studied from an automotive crashworthiness point of view. These sandwich structures have shown superior performance in terms of absorbing impact energy (Syed Quadri et al. 2017). The flexural fatigue behavior of S2 glassfibre/vinyl ester sandwich composites with two diverse PVC cores and densities showed that the fatigue or endurance strength increased with the PVC core densities. The life of composites increased along the increasing frequency, but the time to failure showed the opposite trend (K. Kanny and Mahfuz 2005). Sandwich constructed from aramid honeycomb as a core and carbon/epoxy face sheets showed strength degradation is observed in flexural fatigue loading at different stroke and frequency (Coskun and Turkmen 2012). Flexural fatigue tests conducted on CFRP woven/Nomex honeycomb composite at various temperatures and load levels. The variation of the stiffness degradation with the loading levels is shown in (Figure 12.10). Fatigue behaviour of CFRP/Nomex-honeycomb sandwich composite is done under four-point bending at different temperatures 25, 50, 70 and 100° C, which showed stiffness is gradually decreased as temperatures increased (A. Rajaneesh, et al. 2018).

The fatigue behaviour of sandwich composites depends on honeycomb cell height, cell thickness and skin thickness. The aluminum alloy 3003 honeycomb core and Al 5754 skin sandwich composites were tested under buckling and three-point bend fatigue loads. In buckling fatigue, the fatigue strength depends on cell height. Higher cell height (15 mm, 20 mm) showed lower fatigue life and cell height of 10 mm showed higher fatigue life. In three-point bend fatigue, the fatigue life of composites increased by increasing the thickness of the honeycomb material. Also, increasing the cell diameter and decreasing the fatigue strength of composites were both observed in fatigue loading. The effect of skin thickness on fatigue life is only observed in three-point bend loads (Tolga Topkaya, 2016). In the flexural fatigue of Nomex core honeycomb, the morphology of fatigue failure was influenced by the orientation of the core. The fatigue life of sandwich structure is ruled by the life of

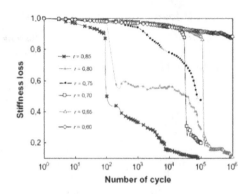

FIGURE 12.10 Stiffness loss of CFRP/Nomex-honeycomb in fatigue loading.

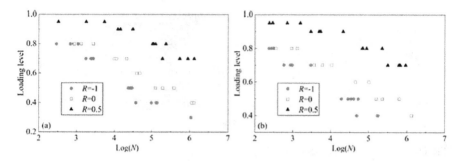

FIGURE 12.11   Loading Level vs Cycles: (a) 1.1 mm face sheet; (b) 2.42 mm face sheet in L direction.

the core and the structure have equal shear stresses and carry the same fatigue life. Figure 12.11 showed the fatigue life for 1.1 mm and 2.42 mm face sheet thickness for different stress ratio and different loading levels (Mingze Ma, et al. 2020).

The fatigue failure in honeycomb sandwich composites starts from face yielding at higher load level case and observed delamination in between face sheet and core at lower loads. The final failure is because of compression of face sheet (M. Hussain, et al. 2019).

12.6   DAMAGE BEHAVIOUR UNDER FATIGUE LOADS

The fatigue damage observed in a PVC foam core and fibre-reinforced face sheet under three-point bend was a result of matrix cracking and crazing. The damage progressed as the stress level or number of cycles increased, and extends through the thickness, and final failure was observed on both sides (Gonabadi, et al. 2016). S.C. Sharma et al. 2004 studied fatigue characteristics of polyurethane foam sandwich composites. In fatigue of sandwich construction of epoxy/glass-PU foam-epoxy/glass sandwich, the damage was limited to skin only. This fatigue damage consists of shear cracks which is bound by glass fibre layers. No interface cracks are observed. These PU foam composite structure have high fatigue strength and stiffness degradation. Polyester/glass-PU foam-polyester/glass sandwich have low fatigue strength and stiffness degradation due to their complete broken behaviour at same fatigue loads. Failure in cyclic loading of two different densities (130 and 260 kg/m³) of fibreglass /vinyl ester S2 sandwich structures with closed cell PVC foam cores, it is observed that (Figure 12.12) crack paths in the density core of 260 kg/m³ were similar in 3 and 15 Hz frequency. At a density of 130 kg/m³, an observed critical crack length at 3 Hz was significantly higher than at 15 Hz (K. Kanny and Mahfuz 2005).

The major failure mode of foam core sandwich structure is compressive failure and shear failure. Compression failure results in damage initiation, but shear failure is the reason for damage progression. The flexural fatigue performance of the foam core sandwich structure at room temperature, low temperature (−55 °C) and hygrothermal environment (77°C, 85%) showed that the failure modes are nearly equal to each other and failed due to core shear. Fatigue life improves in a cold temperature environment (Peiyan Wang, et al. 2013). Under flexural fatigue load, the dominant

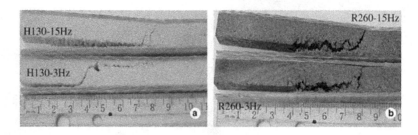

FIGURE 12.12   Fatigue crack path through the core (a) 130 and (b) 260 kg/m^3.

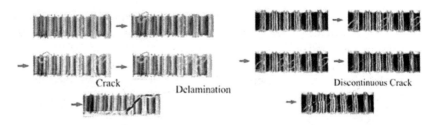

FIGURE 12.13   Fatigue crack growth (a) in L direction (b) in W direction Nomex core honeycomb sandwich.

failure mechanism of the honeycomb sandwich structure is core shear failure. Subsequent delamination in between face skin and core and skin fracture are also observed by failure of the core. The fatigue life of Nomex core honeycomb sandwich is decided primarily on the basis of life of the core. In general, cracks of 45° appear through the thickness in the core of L-direction as shown in Figure 12.13a, and the angle of cracks in the core in the W-direction is smaller, and the cracks are discontinuous as shown in Figure 12.13b (Mingze Ma, et al. 2020).

In the fatigue behaviour of aluminium/aramid fibre core in four-point flexural test, the fracture pattern (Figure 12.14) formation in both W and L configuration are the same and is not affected by the fatigue failure. Both W and L sandwich structure break down in shear with a crack propagating through the core thickness (A. Abbadi, et al. 2010).

The process of fatigue fracture of aluminium core sandwich composites is that the defect is repeatedly created by cracks in the bottom face of the sandwich. From the first load cycle, the next deviation of the vertical cell wall is observed in the middle range. It is observed that (Figure 12.15) crack propagation is stable in zone I (low loads) for the W configuration, while in zone II (high loads) it is observed to be almost stagnant for the L-configuration sandwich structure. Regarding both configurations of failure observed, L configuration is preferable (S. Belouettar, et al. 2009).

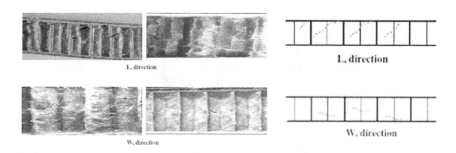

FIGURE 12.14   Fatigue failure modes of the aramid fibre cores in the W and L-directions.

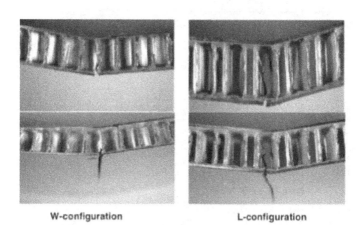

W-configuration   L-configuration

FIGURE 12.15   Crack observation of aluminium core sandwich in the W and L directions.

12.7   CASE STUDY: EFFECT OF SPAN LENGTH AND FIBRE ORIENTATION ON SANDWICH STRUCTURE OF AUTOMOTIVE LEAF SPRING

In the automotive industry, the trend towards better alternative materials more suited to current electrical vehicles, trucks and buses increased, in order to get lightweight, high specific strength and high durability with minimum fuel consumption. In automotive suspensions, it has been observed that the range of stress is from 300 to 750 MPa for passenger and heavy vehicles (Vikas Khatkar, et al. 2019). The leaf spring supports the dynamic of the vehicle in order to provide travel comfort to passengers. Many research works have been done on leaf springs using steel, fibre-reinforced polymer composites and sandwich composites to focus on their improvements on strength, rigidity, weight reduction and fatigue life.

The case study focuses on the effect of span length and fibre orientation on sandwich composite leaf spring under bending load. In numerical modeling using ANSYS, the leaf spring is considered as a rectangular plate of 13 mm thickness and 50 mm width. The length varys as related to span length. The numerical study was carried out for span length as 30 mm, 50 mm, 100 mm, 150 mm, 200 mm and 250 mm respectively. Figure 12.16 shows one of the three-point bend models having a span length of 250 mm.

In sandwich construction of leaf spring (Figure 12.17), the aluminium alloy Al 3003 is core material and glass-fibre reinforced composite (GFRP) is a face sheet material. The core thickness is 10 mm and CFRP sheet is model with 0.1 layer in sandwich construction (Table 12.7).

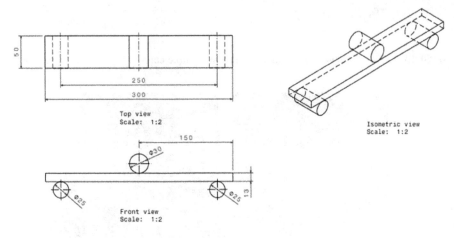

FIGURE 12.16   Three-point bend model of sandwich leaf spring.

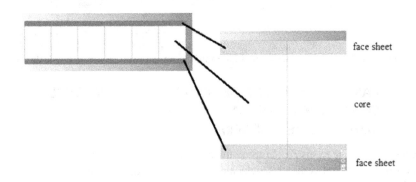

FIGURE 12.17   Composite structure of Al 3003 core and GFRP face sheet.

TABLE 12.7
Mechanical Properties of Core and Face Sheet

Materials	Young's Modulus (GPa)	Poisson's Ratio	Tensile Strength (MPa)	Reference
Al-3003 Core	68.9	0.33	131	(Laurent Wahl, et al. 2014)

Material	E_x GPa	E_y GPa	v_{12}	G_{12} (GPa)	X_T, Y_T (MPa)	X_C, Y_C (MPa)	S (MPa)	Reference
CFRP Face Sheet	3.4	83.5	0.05	6.8	1008	953	125	(Laurent Wahl, et al. 2014)

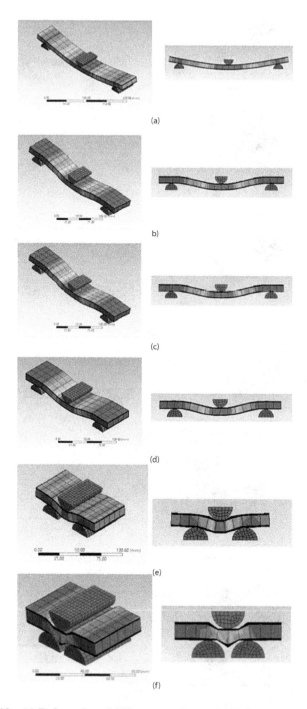

FIGURE 12.18 (a) Deformation of 250 mm span length (b) Deformation of 200 mm span length (c) Deformation of 150 mm span length (d) Deformation of 100 mm span length (e) Deformation of 50 mm span length (f) Deformation of 30 mm span length.

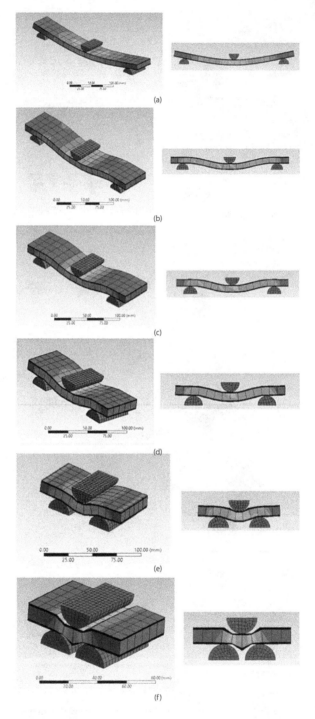

FIGURE 12.19   (a) Stress of 250 mm span length (b) Stress of 200 mm span length (c) Stress of 150 mm span length (d) Stress of 100 mm span length (e) Stress of 50 mm span length (f) Stress of 30 mm span length.

A quasi-static structural analysis of composite structure consists of loading as applied of 3000 N at the middle of the top face. Figure 12.18 shows a three-point bend model of sandwich composite with boundary condition and loading. Figure 12.18a to Figure 12.18e shows the total deformation of leaf spring sandwich composite model with different span length.

Figure 12.19a to Figure 12.19e shows the equivalent von-mises stress of leaf spring sandwich composite model with different span length (Table 12.8 and Table 12.9, Figures 12.20 and 12.21).

12.7.1 FIBRE LAY-UP ORIENTATION

TABLE 12.8
Total Deformation for All Lay-Up Orientation Along with Different Span Length

Span Length (mm)	Deformation (mm)					
	(0/90/0/90)	(0/0/0/0)	(0/45/0/45)	(0/45/90/0)	(45/60/45/60)	(30/45/60/30)
30	3.3318	3.3791	3.3525	3.3329	3.3256	3.3342
50	3.0435	3.1081	3.0631	3.0747	3.0443	3.0635
100	3.1174	3.1179	3.1205	3.1211	3.129	3.1265
150	3.0541	3.049	3.0524	3.055	3.0603	3.0574
200	3.0305	3.0276	3.029	3.0308	3.0334	3.0316
250	3.0169	3.0116	3.0122	3.0131	3.014	3.0132

TABLE 12.9
Equivalent (von Mises) Stress for All Lay-Up Orientation Along with Different Span Length

Span Length (mm)	von Mises stress (MPa)					
	(0/90/0/90)	(0/0/0/0)	(0/45/0/45)	(0/45/90/0)	(45/60/45/60)	(30/45/60/30)
30	21567	21557	21582	21725	21758	21734
50	3624.6	3282.5	3483	5095.5	4785.2	5044.2
100	1805.2	1463.5	1499.3	1968	2013.4	1895.9
150	876.91	649.53	729.54	1051.1	1003.3	926.17
200	568.65	353.22	394.07	1052.07	598.89	586.3
250	157.9	197.15	206.42	385.27	327.44	324.26

Span Length vs Von-Mises Stress for All Layup

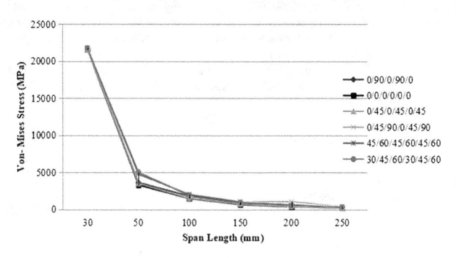

FIGURE 12.20 Span length vs. total deformation all lay-up direction.

Span Length vs Deformation for All Layup

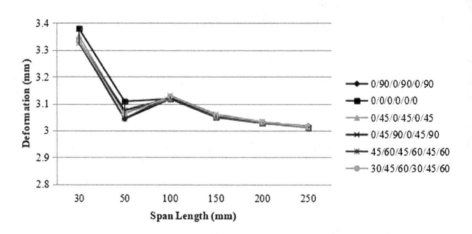

FIGURE 12.21 Span length vs. total deformation all lay-up direction.

12.8 SUMMARY

The study of sandwich core materials and their structure as used for automotive applications provides clear information on sandwich composite core and face sheet or skin material for the preparation of the composite structure. Review study of various experimental tests and numerical studies verified the static and fatigue strength and damage behaviour of the sandwich structure. The tests are carried out under different loading conditions with different configurations of composite structures. Based on this quasi-static and fatigue failure behavior, several parameters that influenced the properties of the sandwich material were studied. Static and fatigue damage behavior correlates with a better understanding of sandwich composite failure and motivation to create new sandwich structures that will be used in numerous automotive utilisations in the future. The newly sought sandwich structure must be suitable for all composite materials–related automotive applications and must be suitable for the end user.

REFERENCES

A. Abbadi, Z. Azari, S. Belouettar, J. Gilgert, P. Freres, "Modelling the Fatigue Behaviour of Composites Honeycomb Materials (Aluminium/Aramide Fibre Core) Using Four-Point Bending Tests". *International Journal of Fatigue* 32 (2010): 1739–1747.

Mohd Zahid Ansari, Sameer Rathi, Kewal Chand Swami, Sunil, Sonika Sahu. "Mechanical Behaviour of Polymer Sandwich Composites under Compression". *American Journal of Materials Science.* 5(3C) (2015): 107–111.

J. Arbaoui, Y. Schmitt, J. Pierrot, F. Royer. "Effect of Core Thickness and Intermediate Layers on Mechanical Properties of Polypropylene Honeycomb Multi-Layer Sandwich Structures". *Archives of Metallurgy and Materials* 59 (1) (2014): 11–16.

S. Belouettar, A. Abbadi, Z. Azari, R. Belouettar P. Freres, "Experimental Investigation of Static and Fatigue Behaviour of Composites Honeycomb Materials Using Four Point Bending Tests". *Composite Structures* 87 (3), (2009): 265–273.

Keskes Boualem. "Static And Fatigue Characterization of Nomex Honeycomb Sandwich Panel". *Pamukkale University Journal of Engineering Sciences.* 19(7) (2013): 287–292.

Wahid Boukharouba, Abderrezak Bezazi, Fabrizio Scarpa, "Identification and Prediction of Cyclic Fatigue Behaviour in Sandwich Panels". *Measurement* 53 (2014): 161–170.

Sukmaji Indro Cahyono. "Light-Weight Sandwich Panel Honeycomb Core with Hybrid Carbon-Glass Fibre Composite Skin for Electric Vehicle Application". *AIP Conference Proceedings.* 1717 (1), (2016): 040025-1–040025-2.

K. Chemami, J. Bey, Z. Gilgert, B. Azari. "Behaviour of Composite Sandwich Foam Laminated Glass/Epoxy Under Solicitation Static and Fatigue". *Composites Part B Engineering* 43 (2012): 1178–1184.

Onur Coskun and Halit S. Turkmen, "Bending Fatigue Behaviour of Laminated Sandwich Beams". *Advanced Materials Research.* 445 (2012): 548–553.

Amarnath Donga. "Application Of Sandwich Beam in Automobile Front Bumper for Frontal Crash Analysis". *Master Thesis, Wichita State University.* 2011.

Ismaeil Eynali, Seyed Alireza Seyed Roknizadeh, "Numerical Study of the Effect of Temperature Changes on the Failure Behavior of Sandwich Panels with Honeycomb Core". *Journal of Applied Dynamic Systems and Control.* 2 (2) (2019): 1–10.

Hassan Izadi Gonabadi, Adrian Oila, and Steve Bull. "Fatigue of Sandwich Composites in Air and Seawater". *Journal of Bio-and Tribo-Corrosion* 2 (10), (2016): 1–7.

Yash Gupta, Varghese Paul V, Ashish Jacob and Akash Mohanty. "Effect of the Core Thickness on The Flexural Behaviour of Polymer Foam Sandwich Structures". *IOP Science Notes.* (2020): 024404.

Deniz Hara, Gokhan Ozgen. "Investigation Of Weight Reduction of Automotive Body Structures with the use of Sandwich Materials." *Transportation Research Procedia.* 14 (2016): 1013–1020.

Muzamil Hussain, Rafiullah Khan, Naseem Abbas. "Experimental And Computational Studies on Honeycomb Sandwich Structures Under Static and Fatigue Bending Load". *Journal of King Saud University - Science.* 31 (2) (2019): 222–229.

Cihan Kaboglu, Long Yu, Iman Mohagheghian, Bamber R. K. Blackman, Anthony J. Kinloch, and John P. Dear. "Effects of the Core Density on the Quasi-Static Flexural and Ballistic Performance of Fibre-Composite Skin/Foam-Core Sandwich Structures". *Journal of Materials Science* 53 (2018):16393–16414.

K. Kanny, H. Mahfuz, "Flexural Fatigue Characteristics of Sandwich Structures at Different Loading Frequencies". *Composite Structures.* 67 (2005): 403–410.

Kicki F. Karlsson and B. Tomas Astrom, "Manufacturing and Applications of Structural Sandwich Components". *Composites Part A Applied Science and Manufacturing.* 28A (1997): 97–111.

Khatkar, Vikas, R. N. Manjunath, Sandeep Olhan, and B. K. Behera "Potential of Textile Structure Reinforced Composites for Automotive Applications". In *Advanced Functional Textiles and Polymers* (eds S. ul-Islam and B. Butola. (2019): 65–98.

Nitin Kulkarni, Hassan Mahfuz, Shaik Jeelani, Leif A. Carlsson. "Fatigue Crack Growth and Life Prediction of Foam Core Sandwich Composites Under Flexural Loading". *Composite Structures* 59 (2003): 499–505.

Jinga Lin, Zhihua Wang, Longmao Zhao. "Failure and Deformation Modes of Sandwich Beams under Quasi-Static Loading". *Applied Mechanics and Materials* 29–32 (2010): 84–88.

Mingze Ma, Weixing Yao, Wen Jiang, Wei Jin, Yan Chen, Piao Li, "Fatigue Behavior of Composite Sandwich Panels Under Three Point Bending Load". *Polymer Testing.* 91 (2020):106795.

B. J. Manujesh, Vijayalakshmi Rao, "Fatigue Behavior and Failure Mechanism of PU Foam Core E-glass Reinforced Vinyl Ester Sandwich Composites". *International Journal of Materials Engineering.* 3(4) (2013): 66–81.

McCarthy, C., "14 - Micromechanical failure analysis of advanced composite materials". In *Woodhead Publishing Series in Composites Science and Engineering, Numerical Modelling of Failure in Advanced Composite Materials,*(eds. Pedro P. Camanho, Stephen R. Hallett) Woodhead Publishing, (2015): 379–409.

T. Neumeyer, T. Kroeger, J. F. Knoechel, P. Schreier, M. Muehlbacher and V. Altstaedt. "Thermoplastic Sandwich Structures – Processing Approaches Towards Automotive Serial Production". *21ˢᵗ International Conference on Composite Materials.* 2017.

Haibin Ning, Gregg M. Janowski, Uday K. Vaidya, George Husman. "Thermoplastic Sandwich Structure Design and Manufacturing for The Body Panel of Mass Transit Vehicle". *Composite Structures* 80 (2007): 82–91.

Syed Salman Mohiuddin Quadri, Mirza Shoeb Ahmed Baig, and Naseeb Khan. "Design Analysis and Experimental Evaluation of Sandwich Composites Subjected to Fatigue". *Indian Journal of Science Research* 17(2) (2017): 237–240.

A. Rajaneesh, Z. Yian, G. B. Chai, and I. Sridhar, "Flexural Fatigue Life Prediction of CFRP-Nomex Honeycomb Sandwich Beams". *Composite Structures* 192 (2018): 1–17.

S. C. Sharma, M. Krishna, H. N. Narasimha Murthy, M. Sathyamoorthy, and Debes Bhattacharya. "Fatigue Studies of Polyurethane Sandwich Structures". *Journal of Materials Engineering and Performance* 13(5) (2004): 637–641.

Tolga Topkaya, Murat Yavuz Solmaz, "Fatigue Behavior of Honeycomb Sandwich Composites Under Flexural and Buckling Loading". *ICAMS* 2016. (2016):177–182.

Shubham Upreti, Vishal K. Singh, Susheel K. Kamal, Arpit Jain, Anurag Dixit. "Modelling and Analysis of Honeycomb Sandwich Structure Using Finite Element Method". *Materials Today: Proceedings.* 25(4) (2020): 620–625.

Cagrı Uzay, Necdet Geren, Mete Han Boztepe, and Melih Bayramoglu. "Manufacturing of Light Weight Sandwich Structures by Using Polymer Foam Core". *Istanbul International Conference on Progres in Applied Science.* 2017.

Laurent Wahl, Stefan Maas, Daniele Waldmann, "Fatigue in the Core of Aluminium Honeycomb Panels: Lifetime Prediction compared to Fatigue Tests", *International Journal of Damage Mechanics.* 23(5) (2014):661–683.

Peiyan Wang, Xiaoyu Li, Zhufeng Yue, "Experimental and Numerical Evaluation of Fatigue Behavior of Foam Core Sandwich Structure". *International Journal of Aerospace and Lightweight Structures.* 3(3) (2013): 337–346.

Xiaorong Wu, Hongjun Yu, Licheng Guo, Li Zhang, Xinyang Sun, and Zilong Chai, "Experimental and Numerical Investigation of Static and Fatigue Behaviors of Composites Honeycomb Sandwich Structure". *Composite Structures.* 213 (2019):165–172.

F.P. Yang, Q.Y. Lin, and J.J. Jiang. "Experimental Study on Fatigue Failure and Damage of Sandwich Structure with PMI Foam Core". *Fatigue & Fracture of Engineering Materials & Structures.* 38(4) (2015): 456–465.

Fa Zhang & Ramadan Mohmmed & Baozhong Sun & Bohong Gu. "Potential of Textile Structure Reinforced Composites for Automotive Applications". *Applied Composite Materials.* 20(6) (2013): 1231–1246.

Index

CPSIA information can be obtained
at www.ICGtesting.com
Printed in the USA
BVHW091743190422
634676BV00002B/44